The Ocean's Role in Global Change: Progress of Major Research Programs

Ocean Studies Board
Commission on Geosciences, Environment, and Resources
National Research Council

National Academy Press
Washington, D.C. 1994

The work was sponsored by The National Oceanic and Atmospheric Administrations's Office of Global Programs through the National Science Foundation Grant No. OCE-9313563-R. Such support does not constitute an endorsement of the views in this report by the sponsors.

Cover art by Winslow Homer, titled "Eight Bells." Etching sculpted by John Dois Andrews, the Intaglio Guild. Special thanks to John Morrell.

Library of Congress Catalog Card Number 94-65573
International Standard Book Number 0-309-05043-X

Additional copies of this report are available from:
National Academy Press, 2101 Constitution Avenue, NW, Box 285, Washington, DC 20055.

B-320

OCEAN STUDIES BOARD

COMMISSION ON GEOSCIENCES, ENVIRONMENT, AND RESOURCES

Contents

Preface

The ocean plays a predominant role in regulating both natural and human-induced changes in the planet. The role of ocean circulation and the coupling of the ocean and the atmosphere are basic to understanding Earth's changing climate. Regional events such as El Niño and ocean margin and equatorial upwelling influence climate on both seasonal and longer time scales. The world's population is now large enough to alter the chemical composition of the ocean and atmosphere and to impact the biological composition of Earth.

(*Oceanography in the Next Decade: Building New Partnerships*, National Research Council, Washington, D.C., 1992)

The Ocean Studies Board (OSB) report, *Oceanography in the Next Decade: Building New Partnerships*, highlighted the research programs that contribute to an understanding of the ocean's role in Earth systems. It stressed the importance of developing new partnerships between the federal government and the academic oceanographic community. These partnerships must be based on information transfer among participants. The Ocean Studies Board has strived to accomplish this communication and education in the form of brief summaries of the major research programs, first in 1990, and now at the beginning of 1994.

The OSB published *The Ocean's Role in Global Change: The Contemporary System—An Overview of Major Research Programs* in 1990. It described the major research programs that were ongoing or planned at that time, specifically those designed to study the role of the ocean in short-term climate variability. Since that time, many of the programs described have made substantial progress, and new programs that contribute to the study of global change have been planned and initiated. In addition, the 1990 report did not describe research programs that seek to understand long-term variations ranging from thousands to millions of years—the geological perspective.

The OSB has written *The Ocean's Role in Global Change: Progress of Major Research Programs* to report the progress of the major oceanographic research programs in the past few years. The information necessary for this update was provided by program offices, federal agencies that fund these programs, and the scientists who lead the research efforts, with guidance and overview provided by OSB members. This document does not evaluate or review the research programs. It is intended to serve as an educational reference document for scientists, administrators, managers, Congress, and the public.

The OSB anticipates a follow-up study to look at the out-years of the research programs and anticipate where the next-generation projects relating to the ocean's role in global change should be focused. The OSB also looks forward to working closely with the National Research Council's Board on Global Change as it continues to study the ocean's role in global change.

WILLIAM MERRELL
Ocean Studies Board, *Chairman*

The Ocean's Role in Global Change: Progress of Major Research Programs

Introduction

During the past decade, the debate about whether Earth's climate is changing has intensified. Global—or even regional—climate shifts will have far-reaching implications for world economics, energy utilization, national defense, and the health of terrestrial and marine ecosystems. The potential significance of climate change has made this topic a major issue in national and international policy. In addition to concern by policymakers, the scientific community continues to search for answers to vital questions about the likelihood of climate change and the predictability of its extent and timing. Scientists are developing capabilities to distinguish actual climate change from "noise" due to natural variation of the global climate system on annual and decadal time scales.

Although the role of the ocean in global climate is not fully understood, there is general agreement that it is significant. For example, the concentration of the greenhouse gas carbon dioxide (CO_2) in the atmosphere has been increasing for many decades. However, of the estimated cumulative input of CO_2 from human activities, less than 60 percent is now present in the atmosphere. The ocean is believed to be removing much of the remainder, although the extent of uptake by the terrestrial biosphere also remains a question. Climate models predict that increases in greenhouse gases may lead to significant regional and possibly global climate changes.

The scientific community has initiated large-scale research programs based on studies of the ocean and its relation to global climate and climate related processes. This report, which describes the research programs, is divided into two main sections: programs that study processes that occur over periods ranging from days to hundreds of years—the contemporary system; and those that seek to understand long-term variations ranging from thousands to millions of years—the geological perspective (Box 1). Some programs have both long- and short-term elements. A third section discusses crosscutting data-collection issues important to all programs.

Because of the vastness of the ocean, observations of ocean physical, chemical, biological, and geological characteristics have been relatively sparse. Intense observations are required for process studies, which are focused on the variety of oceanic parameters that interact to produce a set of observed conditions. Processes are sometimes studied by experimentation, for example adding artificial compounds to the ocean to determine how quickly water masses move and mix. Finally, most programs include a modeling component, to attempt to develop predictive relationships between present ocean characteristics and future observations. Models depend on observations and process studies, including information from geological data, in order to assign reasonable values to model parameters and to test model performance. Modeling results can then provide insights to guide further observations and process studies by identifying the variables in the ocean environment that cause the greatest changes in model predictions.

In 1990, the Ocean Studies Board published a brief report, *The Ocean's Role in Global Change: The Contemporary System—An Overview of Major Research Programs*. It described the major research programs that were ongoing or planned at that time, specifically those designed to study the role of the ocean in short-term climate variability. Since that time, many of the programs described have made substantial progress. The chapters that follow are updated summaries of the programs described in the 1990 report, with additional information about new or planned programs that contribute to the study of global change.

Box 1—Focus and Goals of the Major Research Programs

THE CONTEMPORARY SYSTEM PROGRAMS

GLOBAL OCEAN OBSERVING SYSTEM (GOOS)

Focus: Observations needed for prediction of El Niño-Southern Oscillation and detection of global change due to greenhouse warming.

Goal: To provide the oceanic component of the Global Climate Observing System.

TROPICAL OCEAN-GLOBAL ATMOSPHERE (TOGA) PROGRAM

Focus: Studies describing the interactions between the tropical oceans and the global atmosphere, especially the El Niño-Southern Oscillation.

Goal: To model the ocean-atmosphere system for the purpose of predicting its variations.

WORLD OCEAN CIRCULATION EXPERIMENT (WOCE)

Focus: Studies of the surface and subsurface circulation of the global ocean.

Goal: To understand ocean circulation well enough to model its present state, predict its future state, and predict feedback between climate change and ocean circulation.

JOINT GLOBAL OCEAN FLUX STUDY (JGOFS)

Focus: Studies investigating the role of marine organisms and chemistry in modulating global climate change.

Goal: To gain a better understanding of how carbon dioxide is exchanged between the atmosphere and the surface ocean and how carbon is transferred to the deep sea.

GLOBAL OCEAN ECOSYSTEM DYNAMICS (GLOBEC) PROGRAM

Focus: Studies elucidating how changing climate alters the physical environment of the ocean and how this in turn affects marine animals, especially zooplankton and fish.

Goal: To predict the effects of changes in the global environment on the abundance, variation in abundance, and production of marine animals.

ATLANTIC CLIMATE CHANGE PROGRAM (ACCP)

Focus: A combination of studies utilizing historical data, modeling, and direct observation and monitoring of middle and high latitudes of the North Atlantic.

Goal: To understand air-sea interactions between the Atlantic Ocean and the global atmosphere.

ACOUSTIC THERMOMETRY OF OCEAN CLIMATE (ATOC) PROJECT

Focus: Acoustic propagation studies measuring the speed of sound along long distance undersea paths.

Goal: To characterize warming trends in the ocean on global scales.

GLOBAL OCEAN-ATMOSPHERE-LAND SYSTEM FOR SEASONAL-TO-INTERANNUAL CLIMATE PREDICTION (GOALS) PROGRAM

Focus: Studies investigating the variations in sea-surface temperature, soil moisture, sea ice, and snow and the processes that control these conditions.

Goal: To gain a better understanding of global climate change variability on seasonal to interannual time scales for the purpose of predicting this variability.

LAND-OCEAN INTERACTIONS IN THE COASTAL ZONE (LOICZ) PROGRAM

Focus: Studies of fluxes in the coastal zone, how changes in the coastal zone alter the fluxes, and how they will affect the global carbon cycle and trace gas composition of the atmosphere.

Goal: To understand the impacts from changes in climate, sea level, land use, and ecosystem functioning for use in the creation of long-term, sustainable policies for coastal management.

ARCTIC SYSTEMS SCIENCE (ARCSS) PROGRAM

Focus: Paleoenvironmental multidisciplinary studies to address the physical, chemical, biological, and social processes of the Arctic system.

Goal: To understand the processes of the Arctic system in order to predict environmental change on decade-to-century time scales.

THE GEOLOGICAL PERSPECTIVE PROGRAMS

OCEAN DRILLING PROGRAM (ODP)

Focus: Collection and analysis of deep-sea cores from around the world to help reconstruct the paleoceanographic record of past climatic and oceanic conditions.

Goal: To reconstruct the Earth's paleoceanography and more importantly to begin to understand the mechanisms that drive changes in climate and oceanic conditions.

RIDGE INTER-DISCIPLINARY GLOBAL EXPERIMENTS (RIDGE)

Focus: Integrated observational, experimental and theoretical studies to determine the primary processes that have shaped the evolution of our planet, and the long-term temporal variations that may have modified the past climate of Earth.

Goal: To understand the causes and predict the consequences of physical, chemical, and biological fluxes within the global spreading center system.

THE GLOBAL CHANGE AND CLIMATE HISTORY PROGRAM OF THE U.S. GEOLOGICAL SURVEY

Focus: Paleoclimate and current climate processes research addressing environmental change related to increasing human activities

Goal: To provide relevant information on global climate change to the government and the research communities.

MARINE ASPECTS OF EARTH SYSTEM HISTORY (MESH)

Focus: Studies of the paleoceanographic record to address numerous research themes including ocean geochemical and climate change and climate sensitivity and variability.

Goal: To determine the sensitivity of the climate system to natural changes in solar radiation.

The Contemporary System

The depth, breadth, and complexity of the global ocean, covering more than 70 percent of Earth's surface, have challenged our ability to explore, measure, and comprehend its controlling processes and to predict its behavior. Technology has evolved to the point where we can study the ocean on a global scale and study its interactions with the land and the atmosphere. Such studies have gained increased importance because expanding populations and development, and an increase in atmospheric carbon dioxide and other greenhouse gases, will impact the physical, chemical, biological, and geological processes in the global ocean. Interactions among these processes are responsible for the distribution and abundance of plants and animals in the ocean and also produce our climate.

Changes in climate are a critical factor governing life on Earth, and the ocean plays a significant role in climate control. An understanding of short-term, coupled atmosphere-ocean effects like the El Niño-Southern Oscillation (ENSO) can have immediate economic and societal impacts. Agricultural production, and therefore food supplies and economies, is directly affected by climate variations. Changes in climate and the response of the ocean will greatly affect coastal areas resulting in rising or falling sea level, changes in coastal upwelling, and seawater intrusion into freshwater aquifers. We must understand the role of the ocean and sea-ice in climate change, particularly changes in atmospheric gases like carbon dioxide, in order to make predictions and minimize adverse effects on humankind.

The ocean and the atmosphere store heat derived from solar radiation and redistribute the heat from equatorial to polar regions. The thermal inertia of the ocean may reduce the speed of transition from one climate regime to another, as the slow overturning of the deep-ocean limits heat absorption and release at the ocean surface. The ocean is also a major source and sink for gases and chemicals that affect climate, such as carbon dioxide, water vapor, and dimethyl sulfide. Evaporation from the ocean is the main source of water for the global hydrological cycle, in which moisture is distributed by atmospheric motions and returned through precipitation, either directly to the ocean or indirectly as runoff from the continents. The deep circulation of the ocean is strongly coupled to surface processes in polar regions. Most water that flows to the ocean bottom is formed in these regions as surface water becomes colder (through interaction with cold overlying air) and saltier (through ice formation). The sinking of this dense water can have a large effect on global climate because it carries greenhouse gases and heat with it to the ocean bottom, out of contact with the atmosphere for hundreds of years. Formation of deep water in the North Atlantic Ocean can be diminished and reestablished over decadal time scales. There is increasing evidence that anomalously hot summers or cold winters in the United States are related to particular ocean conditions that cause distortions in weather patterns across the country for weeks at a time.

At certain times, developments in basic understanding and technology coalesce to produce dramatic advances. Such a breakthrough is now possible in ocean science with new observational tools and techniques, such as remotely operated vehicles, satellite-borne sensors, trace chemical measurement, acoustic techniques, long-life buoys and floats, and seafloor seismometers. Progress in electronics has provided supercomputers and work stations that can facilitate the processing and analysis of data and the construction of detailed models. The new technology and innovative ideas now available have provided the basis for developing scientific programs on a scale that has not been attainable previously. Successful initiation and completion of the interdisciplinary global experiments described in this report could produce a tremendous increase in our understanding of how Earth works as a system.

The World Climate Research Program (WCRP) began in the early 1980s, using new in situ and satellite measurement techniques. It is aimed at understanding long-term weather variability and climate change. In recognition of the central role of the ocean in these processes, WCRP has focused on understanding the interaction of the ocean and the atmosphere. Programs falling under the aegis of WCRP include the Tropical Ocean-Global Atmosphere (TOGA) program and the World Ocean Circulation Experiment (WOCE). In addition, WCRP will continue to study the processes that control the exchange and transport of energy and water within the global climate system. These programs, for which the United States is providing considerable resources, furnish a context for much of U.S. ocean science research.

The International Geosphere-Biosphere Program (IGBP) is aimed at describing and understanding the interactive physical, chemical, and biological processes that regulate the total Earth system, the unique environment it provides for life, the changes that are occurring in that system, and the manner in which these changes are influenced by human actions. The Joint Global Ocean Flux Study (JGOFS), Marine Aspects of Earth System History (MESH), and Land-Ocean Interactions in the Coastal Zone (LOICZ) programs, which are described in this report, are elements within the IGBP. [The Past Global Climate (PAGES) program is also an IGBP element, though it is not included here.]

Global Ocean Observing System

The Global Ocean Observing System (GOOS) program will initially emphasize those observations needed for prediction of the El Niño-Southern Oscillation (ENSO), the consequent rainfall and temperature patterns, and observations needed for detection of global change due to greenhouse warming, such as absolute sea level and average ocean temperatures. GOOS is planned to have five application modules, including climate, living marine resources, marine weather and operational ocean services, health of the ocean, and the coastal zone. The climate module of GOOS provides the oceanic component of the Global Climate Observing System (GCOS) and includes the efforts related to global long-term climate change. GOOS will be based in so far as possible on operational measurements (i.e., observations made routinely, essentially permanently, and with societal needs in mind). GOOS is led by the Intergovernmental Oceanographic Commission (IOC) in cooperation with the World Meteorological Organization (WMO), the International Council of Scientific Unions (ICSU), and the United Nations Environment Program (UNEP).

Drawing strongly upon the successes of the TOGA program, the main observational systems for GOOS will initially be moored buoys, primarily in the tropical Pacific; volunteer observing ships; drifting buoys; island tide gauges; and satellites to measure sea-surface temperature and determine surface topography. Future instrument systems might include autonomous profilers [e.g., Autonomous Lagrangian Circulation Explorer (ALACE)], acoustic sensing of average ocean temperature, and in situ biogeochemical measurements. Overlaid on the measurement systems will be satellite telemetry, quality control, and data management, all of which exist now but need improvement and expansion. The design and enhancements of GOOS are based on the current research programs described in this report, planned future programs, and close cooperation between the research and operational communities. Several of the programs described in this report will also furnish important contributions to the development and operation of GOOS.

Tropical Ocean-Global Atmosphere Program

The Tropical Ocean-Global Atmosphere (TOGA) program was designed as the first major World Climate Research Program (WCRP) project and concentrates on the study of the ENSO cycle. The Southern Oscillation is a seesaw-like variation, over several-year intervals, of barometric pressure differences between the South Pacific Ocean and the western Pacific/eastern Indian Ocean. El Niño is a manifestation of the warm phase of the cycle, in which the pool of warm water normally observed in the western Pacific Ocean shifts eastward, diminishing upwelling of cold water along the coast of western South America and along the equator and shifting rainfall patterns away from Australia and Indonesia and eastward into the Pacific. The economic impacts of excess rainfall and flooding in South America and droughts in Australia alone are estimated to be in the billions of dollars.

The U.S. TOGA program has been concerned primarily with studies of the ENSO cycle in the tropical Pacific Ocean and its effect on global climate. The observational phase of TOGA started in 1985 and runs through the end of 1994.

The TOGA program was designed to (1) describe the interactions between the tropical oceans and the global atmosphere in sufficient detail to determine the predictability of the global climate system on seasonal to interannual time scales, (2) understand how and why these ocean-atmosphere interactions occur,

(3) model the coupled system for the purpose of predicting its variations, and (4) design a data collection and distribution system sufficient to achieve the first three objectives. The program includes process studies, long-term observations (over several ENSO cycles), and modeling. Modeling efforts aim to provide a quantitative description of ocean and atmosphere characteristics and to explain why and how these characteristics change over time and space. Coupled TOGA models, used for both simulation and prediction, require data on windstress (which drives surface ocean currents) and sea-surface temperature, in addition to the thermal structure of the upper tropical Pacific.

At present, data are collected regularly from the TOGA Observing System consisting of: (1) the TOGA Tropical Atmosphere Ocean (TAO) array—approximately 65 moored buoys that incorporate instruments to measure the surface winds and the thermal structure of the upper ocean and telemeter the information instantaneously to satellites for immediate distribution through the Global Telecommunications System (GTS), where it is used for research and weather prediction; (2) a network of several equatorial moorings that measure the vertical structure of currents; (3) the TOGA sea-level network—sea-level gauges in the Pacific and Indian Ocean; (4) a Voluntary Observing Ship (VOS) network that measures upper ocean temperature from expendable instruments dropped from merchant vessels in all three tropical oceans; (5) a drifting buoy array that measures tropical sea-surface temperatures and near surface currents; (6) a drifting buoy array that measures sea-surface temperature and sea-level pressure over all three tropical southern oceans; and (7) a Trans Pacific Profiler Network consisting of eight radar sites that measure atmospheric wind profiles.

The TOGA Coupled Ocean-Atmosphere Response Experiment (TOGA COARE), a major process experiment in the eastern Pacific, has just been completed (Box 2). This experiment measured processes influencing interactions between the atmosphere and the warm-water pool in the western Pacific Ocean, including measuring the convective processes in both the atmosphere and ocean that influence these interactions.

The TOGA Program on Seasonal to Interannual Prediction (T-POP), a research program, has been instituted to develop the models and methods needed to provide socially useful predictions of aspects of ENSO a month to a year in advance using data provided by the TOGA Observing System (Box 2). Program participants include scientists from the National Atmospheric and Oceanic Administration (NOAA) and National Aeronautics and Space Administration

(NASA) laboratories, the National Center for Atmospheric Research, and several U.S. universities working on coupled dynamic models for prediction. Informal participation by international colleagues is encouraged.

Planning is under way to form a multinational center, the International Research Institute for Climate Prediction (IRICP), with the following goals: (1) to institutionalize and regionalize short-term climate predictions for the benefit of those nations affected by ENSO variations and (2) to train people from those countries to make, understand, and use these predictions for social and economic benefit. The operating concepts embodied in the IRICP are being demonstrated in a pilot project currently in operation at Lamont-Doherty Earth Observatory.

Because the observational phase of TOGA is drawing to a close, a planning process has been under way both nationally and internationally for programs to maintain and expand the TOGA Observing System as appropriate and to use and expand the prediction results of TOGA. Internationally, the World Climate Research Program is planning the Climate Variability and Predictability (CLIVAR) program whose designated Focus 1 will be on seasonal to interannual global variations and predictability. Nationally, the United States is planning the Global Ocean-Atmosphere-Land System (GOALS) program, a contribution to CLIVAR Focus 1. Efforts are also under way to transfer parts or all of the TOGA Observing System, developed initially as a research observing system, to a permanent observing system in support of the Global Ocean Observing System for use in regular and systematic prediction.

The U.S. component of TOGA is a coordinated effort among NOAA, NASA, the National Science Foundation (NSF), and the Office of Naval Research (ONR). Nationally, the TOGA program was managed by the U.S. TOGA Office which coordinated interagency funding and was advised by the NRC/TOGA Advisory Panel. Internationally, the program was managed by the International TOGA Office, funded by the Inter-governmental TOGA Board, and advised by the TOGA Scientific Steering Group.

Box 2—Major TOGA Accomplishments

(1) Creation and maintenance of the TOGA Observing System, consisting of XBTs (Expendable Bathythermographs), drifters, the TOGA TAO array, a current meter array, a sea-level network, and a set of atmospheric sounding radars.

(2) Measurement of the thermal state of the upper ocean and the winds over several ENSO cycles from 1985 to 1995, including the warm phases of 1986-87, 1991-92-93, and the cold phase of 1987-88.

(3) Development of climatologies of sea-surface temperature (SST), currents, subsurface thermal structure, and surface winds through measurements by the TOGA Observing system.

(4) Development of an operational SST product involving satellite observations combined with in situ drifter measurements.

(5) Development of an operational ocean model for assimilating Pacific Ocean observations and creation of gridded data fields for the entire tropical Pacific for use in research and prediction.

(6) Creation of a set of data centers for the quality control, archiving, and dissemination of TOGA data, in particular the Subsurface Data Center in Brest, France; the Upper Air Data Center in Poona, India; and the Sea Level Center in Hawaii.

(7) Development of coupled atmosphere-ocean models for the simulation of ENSO and subsequently development of theoretical ideas about the genesis and evolution of ENSO, in particular the "retarded oscillator mechanism."

(8) Design and conduct of several smaller process studies (Tropic Heat Experiment, Tropical Instability Wave Experiment, Equatorial Mesoscale Experiment) and a major international process experiment (TOGA COARE) involving aspects of ENSO in and over the tropical Pacific.

(9) Demonstration of the ability to predict aspects of ENSO a year or so in advance and establishment of routine and systematic predictions, thereby opening the new age of short-term climate prediction.

(10) Set up the TOGA Program on Seasonal to Interannual Prediction (T-POP) to develop the coupled dynamical models and other elements necessary for the regular and systematic prediction of climate a year or so in advance, initialized with the data provided by the TOGA Observing System.

(11) Fostering a program to reanalyze of the global atmospheric data, thus creating a climate data set so that TOGA coupled models would have an initialization and validation set for climate prediction.

(12) Participation in the design of the International Institute for Climate Prediction, a major multinational institute for predicting aspects of ENSO a year or so in advance and for maximizing the social and economic utility of these forecasts by forging partnerships with countries most in need of these forecasts.

Summary

The original aims of the TOGA program were to investigate the feasibility of predicting interannual variations in the tropics characteristic of ENSO, to design an observing system to understand the ENSO phenomenon, and to initialize predictions of ENSO. A TOGA Observing System has been established in the tropical Pacific to relay surface and subsurface information to the GTS in real time. Data collected with the TOGA Observing System has helped develop coupled atmosphere-ocean models for the simulation of ENSO events. Using these models and data collected continuously by the TOGA Observing System, researchers have demonstrated that there is significant skill in predicting some aspects of ENSO a year or so in advance (the prediction is a noticable improvement over a prediction that relies solely on the seasonal cycle). The first real-time ENSO forecast using coupled models was made in early 1986, and this ability has since been demonstrated by a number of coupled prediction systems. Today, experimental ENSO forecasts are published routinely.

As a result of TOGA, scientists are now close to establishing a regular and systematic climate prediction capability, using coupled models and sophisticated data assimilation systems, and an operational observing network to support this capability. Accomplishments in these areas have been substantial (Box 2). Predictions have been used advantageously by numerous countries, including the United States, Peru, Brazil, and Australia, for agricultural and water resources planning. Although few studies of the economic impact of ENSO events have been carried out, it has been estimated that the ability to predict an El Niño event at least 6 months in advance with a 60 percent probability, could save the U.S. agricultural sector alone between $0.5 and 1.1 billion per event—the average annual savings would be $183 million per year over a 12-year period (Workshop on the Economic Impact of ENSO Forecasts on the American, Australian, and Asian Continents, 1993). Assuming that forecasts continue to become more skillful, the ability to anticipate climate and mitigate its effects in the United States and other nations could result in even larger savings in the agricultural, fisheries, and water resources sectors of the economies.

World Ocean Circulation Experiment

Ocean circulation is related to climate on a decades-to-centuries scale, through the transfer of heat, momentum, and greenhouse gases between the atmosphere and the ocean. Thus, in order to understand and predict global climate change, on these time scales improved understanding of ocean circulation is crucial. To that end, the WCRP established the World Ocean Circulation Experiment (WOCE).

WOCE studies surface and subsurface circulation of the global ocean. The field program began in 1990 and extends through 1997; it is anticipated that the synthesis phase will continue until 2005. The primary WOCE goal is to understand ocean circulation well enough to model its present state, to predict its future state under a variety of assumptions, and to predict feedback between climate change and ocean circulation. This goal will be met by describing (1) present ocean circulation and variability, (2) air-sea boundary layer processes, (3) the role of exchange among different ocean basins in global circulation, and (4) the effect of oceanic heat storage and transport on the global heat balance.

The WOCE program consists of several related parts, the largest of which is a global survey called Core Project 1 (Box 3). This cooperative international project integrates measurements from satellites, voluntary observing ships (VOSs), moorings, subsurface floats, surface drifters, tide gauges, and research vessels. This global hydrographic survey measures (1) water density; (2) various natural tracers, such as salinity, oxygen and nutrients; and (3) man-made tracers of water motions, such as chlorofluorocarbons (CFCs). Subsurface floats and current meter moorings augment the global survey with direct observations of ocean current velocity. Objectives of the survey are to (1) quantify oceanic transport of heat and the pathways of downward water movement by which atmospheric gases are transported into the deep ocean and (2) provide data to model observed circulation patterns. An upper ocean program will focus on the atmosphere-ocean fluxes that drive the ocean, feedback to the atmosphere, and variations in upper ocean temperature and heat storage.

The U.S. contribution to Core Project 1 began in 1991 in the Pacific Ocean and will be completed by mid-1994. Unfortunately, there will likely be a large gap in coverage in the western North Pacific, where commitments of other nations will not be met. Following the Pacific, U.S. attention will turn to the Indian Ocean, where a major international effort is planned to begin in late 1994 and continue into 1996. U.S. Core Project 1 work in the Atlantic, other than expendable bathythermograph deployments that began in 1990, is still undecided but is scheduled to begin in 1996.

Work in the Southern Ocean (Core Project 2) concentrates on the Antarctic Circumpolar Current (ACC), which connects the Atlantic, Pacific, and Indian oceans, and its interaction with the waters to the north and south. This program includes studies of the formation and spread of cold, dense, high-latitude water masses, as well as measurement of surface temperature, pressure, and velocity from surface drifters. Time series and repeat hydrographic measurements in the areas south of America, Africa, and Australia, together with altimetric measurements from satellites, will give insight into the variability of the ACC. Unfortunately, at present there is little chance of long-term absolute measurements of the transports because of the expense associated with setting up large enough mooring arrays.

Core Project 3 focuses on specific processes important to ocean circulation and modeling. The Subduction Experiment (1991-93) examined the process by which surface water is conditioned and mixed downward into the thermocline. The Tracer Release Experiment (1992-93) has provided the first direct open ocean measurements of vertical and lateral diffusivity of a tracer (the inert, anthropogenic substance, sulfur hexafluoride). The Deep Basin Experiment is examining deep and abyssal flow in the Brazil Basin. These three process studies have been carried out principally by U.S. scientists with assistance from scientists from the United Kingdom, Germany, and France. Enhanced sampling of the North Atlantic Ocean (especially with repeat hydrography, floats, and drifters) is planned through international cooperation and joint work with the Atlantic Climate Change Program (ACCP). However, the full suite of planned Core Project 3 studies is not likely to be implemented because of budgetary constraints.

Progress toward WOCE objectives has resulted in part from a series of technological improvements including: (1) improved meteorological observations from ships and buoys, (2) a new accelerator mass spectrometry facility for measuring radioactive carbon (^{14}C), (3) improved methods for extracting and measuring CFC and helium/tritium, (4) a new type of acoustic Doppler current profiler, (5) autonomous pop-up floats that report their position via Argos satellites at intervals of several weeks for periods up to 5 years, (6) better surface drifters fitted with surface pressure sensors, and (7) an automatic XBT launcher. New procedures for quality control and storage of data have been developed. These will ensure that the WOCE data set remains internally coherent and will be of use for many years in the future.

Despite setbacks, there has been considerable progress toward a better description of general ocean circulation and improvements in modeling ocean circulation. The incorporation of CFC, helium/tritium, and ^{14}C sampling into the hydrographic program is providing a global ocean inventory of these tracers for the first time—an important environmental baseline for the global change research community. Also, the first heat transport analyses addressing the role of the global ocean are being completed currently and will be refined as WOCE proceeds. There has also been steady progress toward improved surface specification of boundary conditions and model-based estimates of surface flux. Major advances have been made in our ability to establish a truly global model of the ocean, with realistic time and space scales. U.S. support for the program comes from NSF, NOAA, ONR, NASA, and the Department of Energy (DOE).

Box 3—Major WOCE Accomplishments

(1) Initiation of global-scale in situ measurements of mid-depth circulation following the successful development and testing of free-drifting ALACE floats and advances in float technology.

(2) Determination of full-depth currents in the equatorial Pacific region by use of lowered Acoustic Doppler Current Profilers (ADCPs).

(3) Direct long-term measurements of the variability in formation rate and transport of North Atlantic Deep Water in the North Atlantic Ocean.

(4) The most complete and accurate description of water masses in the Pacific and South Atlantic oceans as a result of the WOCE Hydrographic Program.

(5) Initiation of measurements to provide the first global inventories of chlorofluorocarbons, helium, tritium, and carbon dioxide.

(6) New estimates of the vertical diapycnal diffusivity during both summer and winter in the central North Atlantic.

(7) An intensive study of the process of subduction in the upper ocean thermocline.

(8) Improved air-sea flux estimates in models resulting from, for example, the inclusion of satellite surface wind-speed data and real-time ice cover data, better stratus cloud parameterization, and the use of an improved spectral statistical interpolation objective analysis system.

(9) New estimates of the meridional heat flux across mid-latitudes in the South Pacific and North and South Atlantic, suggesting an imbalance in heat transport between the northern and southern hemispheres.

(10) Improved fine-scale global ocean models with more realistic physics and bathymetry, driven by real data rather than climatology.

(11) Establishment of data assembly centers and their associated quality control methods as a basis for better data management and data sharing.

(12) Contributions to the analysis of satellite-altimeter data from the Ocean Topography Experiment (TOPEX/Poseidon) and the European Remote Sensing (ERS-1) satellite missions, leading to an estimated orbit uncertainty of 8-9 cm with a geographically correlated component of better than 3 cm and to an improved surface wind field from scatterometer measurements.

(13) Establishment of a facility for measuring radiocarbon in small volumes of sea water and other matrices using Accelerator Mass Spectrometry.

(14) Development of a multiple XBT launcher for high-resolution XBT deployments on Voluntary Observing Ships.

Summary

The WOCE program's primary purpose is to model long-term climate change. Decade-long variations in ocean climatology that affect economic activities such as fisheries and agriculture are known, but the causes of these changes are not understood. The first goal of WOCE is to develop models to predict climate change on decadal scales and to collect the data necessary to test them. WOCE is already making significant strides in this area (Box 3). However, field programs to date have concentrated primarily on the Pacific and South Atlantic oceans. U.S. work in the Indian Ocean is scheduled to begin only in late 1994. A major initiative for the North Atlantic is proposed to begin in 1996, and studies of variability elsewhere will continue for some time. WOCE's second goal is to determine whether specific WOCE data sets are representative of the long-term behavior of the ocean and to find methods for quantifying decadal changes in ocean circulation. This goal will be fulfilled only if appropriate resources are supplied to long-term programs such as climate variability and predictability and global ocean observing system. The research conducted under WOCE is providing indications of what should be measured and where such measurements should be taken to monitor ocean circulation. Process studies are providing new views on small-scale oceanographic processes

Box 4—Future Plans for WOCE

(1) Completion of an integrated study of the Indian Ocean beginning in late 1994.

(2) Enlargement of global drifter and float data sets and improved measurements of deep flow from current meter deployments.

(3) Complete process studies in the Atlantic Ocean.

(4) Launching of the advanced Earth Observing Satellite (NSCATT/ADEOS) plus other satellites to improve global wind field determination.

(5) Continued improvements in modeling, including methods for data assimilation.

(6) Implementation of a major synthesis phase from 1998 through about 2005 to extract the maximum information from the WOCE data sets.

(7) Transfer of pertinent information to those planning of the ocean component of the GCOS.

(8) Transfer of selected long-term measurements (e.g., upper ocean temperature and salinity from VOSs) to a follow-on global change research program or to the GOOS.

such as vertical diffusivity and subduction, and processes within deep-ocean basins. Additionally, WOCE research is leading to continued improvements in global ocean modeling as scientists begin to use the largest computers at fine model scales. While providing new insights into ocean circulation, these data and models will also form the large-scale physical framework for studies of chemistry and biology being carried out through programs such as Joint Global Ocean Flux Study (JGOFS) and Global Ocean Ecosystem Dynamics (GLOBEC). Following the WOCE field program, a synthesis phase is being planned to provide improved operational models of the global ocean and atmosphere as well as a new understanding of how the ocean works in terms of large-scale circulation. Specific products will include: (1) an ocean climatology for the

WOCE period; (2) estimates of ocean variability such as Ekman transport, meridional heat, mass and property transports, interbasin transport, and ocean surface fluxes and fields; and (3) a mid-depth ocean reference level with horizontal velocity estimates.

Joint Global Ocean Flux Study

Observations demonstrating a continuing increase in the concentration of atmospheric CO_2, linked with numerous model-based forecasts, have led the nations of the world to consider controlling of CO_2 emissions. Since the Industrial Revolution, humankind has been altering the partitioning of carbon in the Earth system and hence altering environmental biogeochemical cycles. Unfortunately, scientists cannot accurately assess the relative roles of the land and ocean system in taking up atmospheric CO_2. Neither do scientists understand the ocean's role in moderating atmospheric carbon dioxide and other radiatively active gases well enough to quantify its impact and to predict future atmospheric concentrations.

The need to understand biological utilization, regeneration, transport, and sediment burial of bio-active elements such as carbon, nitrogen, and phosphorus in the ocean has led to a number of studies and advances in measurement capability in the 1970s and 1980s. The capability for global satellite measurement of ocean color, which is related to the abundance of phytoplankton in surface waters, has provided new impetus to these studies. In situ techniques have also been developed for direct measurements of the amount of sinking organic debris in the water column, as well as high-precision measurements of chemicals present in minute concentrations. The new understanding of ocean biology and chemistry, and the capabilities provided by new techniques, led to the development of the international Joint Global Ocean Flux Study (JGOFS).

The U.S. JGOFS program was initiated in 1984 and the international JGOFS was established by the Scientific Committee on Oceanic Research (SCOR) in 1987. JGOFS provided the first coordinated planning for studies of biogeochemical cycles in the ocean. A major goal of JGOFS is to gain a better understanding of how carbon dioxide is exchanged between the atmosphere and the surface ocean and of how calcium carbonate and organic debris are transferred to the deep sea. In the deep sea, carbon—originally derived from carbon dioxide in the atmosphere—is removed from the short-term carbon cycle.

Objectives of the JGOFS program are to (1) understand the global-scale processes that control carbon, nitrogen, oxygen, phosphorus, and sulfur exchanges in the ocean over time; (2) understand how, and at what rates, gases, salts, and water are transferred within the ocean and across the boundaries between the ocean and the atmosphere, seafloor sediments, and the continents; (3) determine how well productivity can be measured remotely by satellite- or aircraft-borne sensors; (4) determine the dependence of particle production and sinking on physical, chemical, and biological processes; and (5) develop mathematical simulations that predict chemical transfers within the ocean and across ocean boundaries and the effects of these transfers on global environmental changes. The program is using several approaches for data collection, including research vessels, satellites, aircraft, and deep-ocean moorings.

The first JGOFS process study, of the spring phytoplankton bloom in the North Atlantic, was conducted by a group of nations in the spring and summer of 1989 (Box 5). The second process study was carried out in the equatorial Pacific Ocean in 1991 and 1992. Process studies are scheduled for the Arabian Sea (1994-95) and the Southern Ocean (1994-96). In addition to process studies, U.S. JGOFS maintains two long time-series stations located near Hawaii and Bermuda (Box 5). Data collected at these sites on regular ship visits and permanent moorings will allow JGOFS scientists to assess the variability of the ocean environment on a range of time scales. U.S. JGOFS has a global survey component that presently has two major components: (1) a large-scale survey of oceanic CO_2 parameters measured in collaboration with the WOCE Hydrographic Program series of cruises and (2) a large-scale survey of surface ocean chlorophyll/productivity by observations of ocean color from the new Sea-Viewing, Wide Field Sensor (SeaWiFS) due to be launched in 1994.

U.S. federal agencies have coordinated their involvement in U.S. JGOFS experiments as part of the biogeochemical dynamics section of the U.S. Global Change Research Program. NSF, NOAA, NASA, ONR, and DOE are all involved with U.S. JGOFS. JGOFS also has a liaison with WOCE. JGOFS is now a fully international effort and has been designated as a core project of the IGBP with WOCE and GLOBEC.

Box 5—Major JGOFS Accomplishments

(1) Planning and implementation of two oligotrophic time-series stations near Bermuda and Hawaii. A suite of JGOFS core measurements have been made at these locations since 1988, providing new information on the seasonal and interannual variability of ocean carbon system parameters in the oligotrophic Atlantic and Pacific.

(2) Planning and implementation of two major Process Studies in the North Atlantic in 1989 and the Eastern Equatorial Pacific in 1992. The North Atlantic study provided insights into how open ocean phytoplankton blooms export carbon from the surface ocean. The Equatorial Pacific study is producing carbon cycle data under both El Niño and non-El Niño conditions in an area of the world ocean that contributes in a substantial way to the total oceanic carbon budget.

(3) Development of an improved understanding on the nature and distribution of dissolved organic carbon in the ocean.

(4) Accumulation of a large scale data set focused on the air-sea exchange of carbon dioxide.

(5) Continuation of research to understand the role of micronutrients (e.g., iron) in controlling open ocean primary productivity. A cruise in the Pacific in fall 1993 to test several hypotheses of this process.

(6) Elucidation of the scale of oceanic primary production through the Coastal Zone Color Scanner (CZCS) satellite data set and the upcoming data stream from the seaWiFS instrument.

(7) Development of improved ocean models of CO_2 distribution.

(8) Increased acceptance of the concept of microbial cycling and mechanisms of organic matter regeneration.

(9) Development of a series of sensors for continuous observations of such key carbon system parameters as CO_2.

(10) Planning and implementation of the SeaWiFS program—due to have a new ocean color sensor flying by 1994.

Summary

The JGOFS program seeks to quantify the oceanic fluxes of bio-active elements in particular carbon. An increases understanding of carbon, including exchanges with the atmosphere, will reduce the uncertainties associated with the ocean's role in the global carbon cycle and will provide improvements in coupled models which permit prediction of the carbon system in the future. It is important to understand how the ocean carbon cycle responds to perturbations of global temperature. This knowledge will be of immense value to policymakers trying to craft economic and environmental policies based on the best scientific knowledge available.

A list of JGOFS accomplishments (Box 5) reveals progress toward reaching these goals. The time-series studies at Bermuda and Hawaii provide a sense of the range of seasonal and interannual variability in carbon cycle processes characteristic of the vast oligotrophic areas of the world oceans. The various process studies have provided and will continue to reveal details of key carbon cycle controls in globally significant ocean areas. The large-scale surveys of ocean color and carbon dioxide system parameters will reduce our uncertainty about the global atmosphere-ocean fluxes of carbon and their temporal variability.

Global Ocean Ecosystem Dynamics

The goal of the Global Ocean Ecosystem Dynamics (GLOBEC) program is to predict the effects of changes in the global environment on the abundance, variation in abundance, and production of marine animals. The program aims to understand the fundamental physical and biological mechanisms that determine how marine animal populations vary over time and space, with an emphasis on discovering how changing climate alters the physical environment of the ocean and how this in turn affects marine animals, especially zooplankton and fish. The GLOBEC approach is to determine which fundamental physical and biological oceanographic processes control populations, how the controls operate, and how variations in the abundance of organisms can be attributed to natural or anthropogenic causes. Finally, once the linkages among atmospheric forcing and physical and biological processes are elucidated, this understanding can be translated into assessments and predictions of the impact of climate change on marine ecosystems. These goals will be accomplished through an interdisciplinary effort involving physical and biological oceanographers and fisheries biologists, who will conduct modeling studies, field investigations, retrospective data analysis and long-term observations.

The abundance of marine organisms is related in part to factors other than man's influence. For example, sardine and anchovy abundances, estimated from the abundance of scales in layered sedimentary records, have varied by factors of 3 to 10 over the past 1,700 years—long before intense exploitation by man. Phytoplankton and zooplankton also display substantial year-to-year and longer-term variability, suggesting that the physical environment controls populations to some extent. For example, northerly winds in the North Atlantic intensified from the 1950s to the 1970s and were correlated with a decrease in the biomass of phytoplankton and zooplankton. The decline in phytoplankton might have been due to dilution in their concentration that resulted from deeper wind-driven mixing in the euphotic zone. In turn, this may have resulted in poorer feeding conditions for zooplankton and higher trophic level organisms. Thus, relatively small changes in the physical environment can cascade throughout marine food webs.

The early phases of U.S. GLOBEC concentrated on (1) the development of coupled physical and biological models, (2) the development and application of improved sampling and measurement systems, and (3) the planning of site-specific process studies. Toward that end, several modeling investigations and projects to develop molecular techniques were funded, and a process study on the Georges Bank ecosystem in the northwest Atlantic Ocean was planned (Box 6). U.S. GLOBEC is funding physical and biological research (including modeling) retrospective data analysis and process research in the Georges Bank ecosystem. Key processes that will be investigated include: stratification (water column stabilization) and its relation to feeding success, to growth, and to survival of zooplankton and fish larvae; episodic exchanges of water and organisms with the coast due to storms and interactions with warm core rings; and cross-frontal exchanges of plankton and nutrients. The goal is to obtain an understanding of how physical processes control vital rates (e.g., birth, growth, and survival) of key zooplankton and fish species in the Georges Bank ecosystem particularly and in retentive gyre banks generally. In this way, predictions can be made of the biological and ecological impact of likely climate change scenarios.

U.S. GLOBEC is continuing to develop plans for a study of ocean physics and biology off the West Coast of the United States. Planning activities for investigations in the California Current system have identified four components of study. First, a large-scale component to include satellite studies, nested regional numerical modeling, and comparative studies from Baja California to Washington. Second, mesoscale studies that investigate biophysical interactions at the larval stages, with emphasis on processes involving transport, retention, aggregation, and vital rates as functions of location within mesoscale features. Third, retrospective studies that quantify the natural modes of variability in the marine ecosystem, with a specific focus on the linkages between climate, ocean, ecosystem, and population variability. And fourth, modeling activities, including models with various spatial and temporal resolution, data assimilative models, ecosystem structure models, and nested regional and basin-scale models.

U.S. GLOBEC is a component of the U.S. Global Change Research Program and receives funding from NSF and NOAA. In cooperation with GLOBEC-International, U.S. GLOBEC is proceeding with planning for a study of ocean physics, sea-ice dynamics, and marine animal population dynamics in the Southern Ocean.

GLOBEC-International focuses on fundamental scientific issues associated with zooplankton population dynamics and their variability. The stated goal of GLOBEC-International is "to understand the effects of physical processes on predator-prey interactions and population dynamics of zooplankton, and their relation to ocean ecosystems in the context of the global climate system and anthropogenic change." GLOBEC is endorsed by Scientific Committee on Oceanic Research, Intergovernmental Oceanographic Commission, International Council for the Exploration of the Sea (ICES), and Pacific ICES (PICES).

GLOBEC-International has established a research strategy articulated in the GLOBEC Core Program (GCP). The GCP strategy provides a framework by which international and regional programs can join together toward a common goal of understanding zooplankton dynamics in a physical and ecosystem setting. The GCP is evolving into separate but coordinated activities, with working groups meeting throughout the past year to articulate different aspects of the GCP and to prepare for the full implementation of GLOBEC-International. To date, six scientific planning meetings have been completed.

The GCP is being developed in two complementary directions. The general scientific approach is being generated by four working groups: Population Dynamics and Physical Variability, Numerical Modeling, Sampling and Observation System, and GLOBEC Prudence, which will be reviewing historical data to determine its applicability to GLOBEC problems. The results will be applied to specific ecosystems, the other line of GLOBEC investigation.

The development of the scientific approach so far indicates that the GLOBEC-International mission will involve two components. The first involves the population dynamics of zooplankton and is fairly straightforward. The second involves the development of coupled numerical, physical-biological models and observational systems that will involve a significant planning effort and international cooperation. Coupled models and observations are central to determining present oceanic conditions and predicting future conditions. Both have important applications in global-change and fisheries management issues. The modeling-observation systems would attempt to estimate realistic physical and biological fields with mesoscale resolution since these are thought to be the most energetic (and hence, physically variable) and the most biologically demanding.

Box 6—Major GLOBEC Accomplishments

(1) Planning of a major process study of the linkages between climate, ocean circulation, ocean structure, ecosystem structure, and animal population dynamics for the Georges Bank ecosystem in the northwest Atlantic Ocean.

(2) Development of improved regional circulation models for predicting how ocean currents and structure will respond to global climate changes such as greenhouse warming.

(3) Development of an improved understanding of how climate and regional weather impact biological processes such as secondary production and fish recruitment.

(4) Reestablishment of two long-term continuous plankton recorder lines in the northwest Atlantic Ocean that will provide additional data for determining to what extent climate changes have affected biological populations.

(5) Development (through a series of workshops) of an implementation plan for GLOBEC research into the interactions of physical and biological processes in the mesoscale dominated California Current ecosystem to be completed by mid-1994.

(6) Development of a research strategy and a full implementation plan by GLOBEC-International, which will be complete by mid-1994.

Summary

The goal of the U.S. GLOBEC program is to conduct scientific investigation into the linkages among climate, ocean physics, and marine animal populations. If these linkages are understood, then the responses of marine ecosystems to potential future climate changes, whether anthropologically induced or natural, can be predicted, assessed, and better managed. In addition, GLOBEC supports the development of new technologies (i.e., acoustic, optic, and molecular) that promise higher resolution or more cost effective ways to measure the structure and condition of ocean ecosystems and the status of living marine resources.

The first GLOBEC field programs have just begun and accomplishments to date are preliminary (Box 6). A pilot study of the role of water column stratification on the feeding and behavior of larval cod and haddock on Georges Bank suggests that vertical migratory behavior plays an important role in retaining the larval stages of these commercially important species in favorable environments for growth, survival, and recruitment. Eventually, U.S. GLOBEC research on groundfish species (cod, haddock) on Georges Bank may provide the scientific underpinnings for a rational rebuilding of these historically important commercial stocks and of the fishing industry that is dependent upon them.

Finally, U.S. GLOBEC emphasizes the development of coupled physical and biological diagnostic models to understand existing conditions, with the goal of then using them as prognostic tools to assess potential conditions and responses in the future. These prognostic tools will permit better management of marine ecosystems providing stability, profit, and sustainability.

Atlantic Climate Change Program

The Atlantic Climate Change Program (ACCP) is aimed at understanding large scale air-sea interaction between the Atlantic Ocean and the global atmosphere. Much of the early emphasis of ACCP is directed at middle and high latitudes of the North Atlantic for several reasons. First, the subarctic North Atlantic is the only site in the northern hemisphere where convection in the ocean extends from the surface to deeper levels, providing a mechanism for creating persistent sea-surface temperature anomalies one decade to century time scales. Second, there is a good correlation between sea-surface temperature in the northwestern Atlantic and atmospheric surface temperatures averaged over the entire northern hemisphere. For example, the "dust bowl" in the western United States in the 1930s was accompanied by pronounced warmth over the northern North Atlantic. Similarly, the relatively cool climate of the late 1960s and early 1970s was associated with very low sea-surface temperature in the western Subarctic North Atlantic. Finally, in the context of assessing global climate change due to anthropogenic causes, understanding the very low frequency, apparently natural variability of climate will help answer the question: Can we discriminate natural climate variability of very long time scale from the possible effects of human-induced greenhouse warming?

To study these phenomena, ACCP has adopted a three-pronged approach—analysis of historical data, modeling, and direct observation and monitoring of the ocean. The first step in ACCP has been to assemble and examine Atlantic Ocean data collected in the past. The second element of the program is an attempt to validate a whole hierarchy of atmospheric and ocean models using these data. The model hierarchy is intended to range from simple conceptual models to numerical models that link the global ocean and atmosphere. These models are aimed to better understand high-latitude air-sea interactions and to help design an effective system for monitoring low-frequency changes in water mass properties and in the heat balance, which may be linked to persistent sea-surface temperature anomalies. The final element of the program will be to test and deploy instruments for monitoring winds, sea surface temperature, sea ice, and water mass properties and to combine that information with other measurements from satellites, drifting instruments, and ships of opportunity. ACCP will coordinate closely with WOCE in North Atlantic studies.

ACCP analysis of historical data has indicated two types of low-frequency climate variability over the North Atlantic. One type has a period of about a decade with marked observational signatures in surface winds, sea ice coverage, and sea-surface temperature. A second type has a period of four to six decades and appears to be associated with polar-amplified climate variations affecting the entire northern hemisphere. Its observational signature is in the ocean temperature and salinity fields, both at the surface and at depth. The accompanying atmospheric anomalies are weaker than those of the decadal mode and show a markedly different structure. Why this difference should exist between the decadal and the multidecadal time-scale climate fluctuations is an open question being investigated by ACCP modelers.

An abrupt change in the climate regime of the Atlantic Ocean occurred in the late 1960s and early 1970s. This was associated with the formation of a very large patch of low salinity water in the northwest Atlantic, which has been dubbed "The Great Salinity Anomaly." This anomaly drifted off to the east after a few years, but while it was in the Labrador Sea it apparently caused a major disruption of normal wintertime convection in the ocean, leaving a clear signal in surface, as well as deepwater mass properties. Though historical record is incomplete, there is evidence to indicate that a similar event may have taken place around 1910 in the Labrador Sea. It is not clear whether these extreme surface salinity events are associated with the decadal or multidecadal climate variations.

Some of the most important accomplishments of the ACCP have been in the area of modeling. Atmospheric models have shown that the response to high-latitude sea-surface temperature anomalies is much more complicated than, and fundamentally different from, the response to tropical sea-surface anomalies. The chaotic nature of middle and high latitude flows prevent any simple, linear response patterns from emerging, in which cause and effect could easily be identified. This insight explains the confusing results obtained in previous studies of atmospheric response to high-latitude sea-surface temperature patterns.

Ocean models with boundary conditions which mimic the effects of air-sea interaction provide the simplest illustration of how changes in the ocean's thermohaline circulation and in high-latitude ocean convection provide an explanation for low-frequency climate variability in the North Atlantic. Rapid progress has been made in clarifying the early results obtained with these models, which showed that two very different solutions could exist for the same boundary conditions. These results show how the different salinity patterns that may have existed during the last Ice Age could lead to a very different, and greatly amplified, type of climate variability than has existed over the past few thousand years.

Perhaps the single most important accomplishment in modeling has been the simulation of multidecadal Atlantic climate variability by the Geophysical Fluid Dynamics Laboratory (GFDL) global coupled ocean-atmosphere model (Box 7). In 1,000-year-long integration, multidecadal variability associated with changes in strength of the thermohaline circulation comes out clearly. The sea-surface temperature anomalies produced by the model also match the decadal time-scale anomalies observed in the North Atlantic. This successful simulation provides an important building block for future ACCP modeling studies and for the design of practical monitoring systems in the Atlantic.

In planning ACCP, it was recognized that the instrumental record is too short to adequately sample decade-to-century time scales. For this reason the program contains an effort to analyze proxy data for the ocean—ice caps on land. These data can be used to constrain the models and to provide a perspective on the limited climate data available in the instrument record, which may already be contaminated by anthropogenic effects. The records from the Greenland ice cap appear to have enormous potential for the study of North Atlantic climate variability.

Plans for field activities in ACCP are guided by the results of the data analysis and modeling elements of the program. Thus, an original strategy of ACCP was to concentrate attention on the poleward transport of heat by the ocean circulation into the northern North Atlantic. In its early stages the field component of the program focused on continuation of long-term monitoring at 24°N in the Atlantic. In coordination with WOCE, a repeat section was made at 24°N in 1992 along with two shorter parallel sections in the vicinity of the western boundary. Measurements extending over nearly a decade are gradually providing details of the boundary flows near the Bahamas and in the Florida Straits. ACCP will coordinate with WOCE to continue monitoring heat transport at 24°N and to extend monitoring to higher latitudes. One of the most valuable data sources for ACCP has been the time series of hydrographic measurements taken at Bermuda and from the weather ships. While the Bermuda time series is being maintained, the weather ships no longer exist. One of the long-term goals of ACCP is to develop the instrumentation to reinstate the time series of temperature and salinity at weather ship sites in the northwestern Atlantic relying on measurements made by WOCE and other programs for the interior of the ocean.

ACCP has achieved a close working relationship with Canadian oceanographers also studying the decade-to-century climate variability in the Atlantic. Phase II of Climate Variability and Predictability, the new program of the WCRP, is focused on decade-to-century climate variability, and many of the early research accomplishments of ACCP have been incorporated in the early planning of CLIVAR. In the future CLIVAR is expected to play a major role in coordinating ACCP efforts with that of a larger international community interested in the role of the Atlantic in climate variability and climate change.

Box 7—Major ACCP Accomplishments

(1) Assembly and examination of Atlantic Ocean data collected in the past.

(2) Finding that the response to high-latitude sea-surface temperature anomalies is much more complicated than and fundamentally different from the response to tropical ocean sea-surface anomalies using atmospheric models.

(3) Simulation of multidecadal Atlantic climate variability in the GFDL global coupled ocean-atmosphere model.

(4) Continuation of long-term monitoring at 24°N in the Atlantic (in coordination with the WOCE program).

(5) Establishment of a close working relationship with Canadian oceanographers studying the decade-to-century variability in the Atlantic.

Arctic Systems Science

The Arctic region has gained a prominent role in the current debate regarding global change. The Arctic comprises a mosaic of precariously balanced ecosystems that interact intimately with climate. Global climate models have shown that the largest temperature changes may occur in the Arctic. In addition the Arctic has been identified as a potentially key source of global greenhouse gases, especially methane. The recognition of the importance and sensitivity of the polar regions in a changing global environment led to the creation of a new program called Arctic Systems Science (ARCSS).

The ARCSS program has two goals: (1) to understand the physical, chemical, biological, and social processes of the Arctic system that interact with the total Earth system and thus contribute to or are influenced by global change and (2) to improve the scientific basis for predicting environmental change on a decade-to-centuries time scale and for formulating corresponding policy options in response to the anticipated impacts of this change on humans and social systems.

Oceanographic research is a crucial component of ARCSS, although the initiative spans terrestrial, marine, and atmospheric research. The marine environment of the Arctic is an interactive system, comprising the water, ice, biota, dissolved chemicals, and sediments. Several key research areas have been identified, including the effects of energy exchange (for example, from wind or the sun) on temperature, salinity, and density distributions in the water column and carbon removal from the atmosphere and surface waters to the deep ocean and sediments via plant material. These goals are shared with WOCE and JGOFS, and in order for them to give a complete global assessment of ocean processes interrelated with climate, WOCE and JGOFS must have the knowledge of high latitudes that can be supplied through the ARCSS initiative. Because of the remoteness and inaccessibility of much of the Arctic Ocean, satellite sensors play a key role in data gathering.

ARCSS has included three components: (1) Paleoenvironmental Studies, (2) Ocean-Atmosphere-Ice Interactions (OAII), and (3) Land-Atmosphere-Ice Interactions (LAII) (Box 8). The Paleoenvironmental Studies component is, in turn, made up of two activities, the Greenland Ice Sheet Project Two (GISP2) and the Paleoclimates of Arctic Lakes and Estuaries (PALE) project.

The ARCSS Executive Committee has also identified research priorities for the future. The current categories of paleoenvironmental studies (GISP2, PALE) and studies of the contemporary environment (OAII, LAII) are expanded to include archaeology and human-environment interactions, respectively. ARCSS is funded by NSF as part of their contribution to the U.S. Global Change Research Program.

Box 8—Major ARCSS Accomplishments

(1) GISP2 completion of drilling for the longest environmental record ever obtained for an ice core and the longest such record that can be retrieved in the northern hemisphere. Ultimately, the core is expected to yield a 250,000-year climate history.

(2) Precise dating and high-resolution, continuous analysis of the ice core material, yielding a detailed view of climate change well into the last glacial period. Specifically, anthropogenic effects are clearly evident during our industrial era and detailed environmental changes are visible during the last glaciation.

(3) Establishment of research sites in poorly sampled regions and in areas particularly sensitive to rapid climate variations in the initial phase of PALE and, refinement of methods to maximize the paleoclimatic signal obtained from sediment cores.

(4) Completion of two oceanographic cruises under OAII, resulting in a wealth of multidisciplinary data. The Northeast Water Polynya Project focused on primary productivity and biogeochemistry in polynya waters. The second study investigated the shelf waters of the Bering, Chukchi, and Beaufort seas.

(5) OAII support for the U.S. Interagency Arctic Buoy Program and initiated modeling efforts.

(6) LAII (initiated in late 1992) development of a detailed implementation plan for the multidisciplinary Flux Study, which will measure the rates and controls of carbon dioxide and methane fluxes along transects spanning a variety of ecosystems in northern Alaska.

Acoustic Thermometry of Ocean Climate Project

The Acoustic Thermometry of Ocean Climate (ATOC) project is designed to characterize global climatic trends in the ocean by measuring the changes in the speed of sound along long-distance undersea paths. It rests on two principles: (i) sound speed increases with temperature, and (ii) acoustic transmissions can be monitored over gyre and basin scale ranges. This makes it possible to form synoptic horizontal temperature averages that are well suited for measuring climate change. Following a successful 1991 demonstration of the viability of acoustic travel time measurements over trans-oceanic paths (the Heard Island Experiment), ATOC was funded in early 1993 to establish a Pacific Ocean network of sound path measurements to test the feasibility of a future global network for monitoring ocean climate trends.

The advantage of using acoustical measurements is one of scale. Timing sound travel across ocean basins removes small-scale (mesoscale) variability in local temperature to reveal large spatial and temporal scale changes. ATOC has developed a plan to install mid- and eastern Pacific sound sources in the deep-ocean sound channel to establish pathways from California to New Zealand. These sources will transmit low-frequency signals to generate precise timing for sound traveling to Navy and ATOC receivers.

In the first 6 months of project activity, ATOC has established a configuration for the planned network, started construction of the sources and receivers, and developed detailed plans for their installation and operation (Box 9). The initial network will begin operational activities in early 1994 to connect North Pacific paths. Trans-equatorial paths will be established by late spring, and the network will become fully operational by late summer. To date, ATOC has met all of its planned engineering milestones and has developed contingency plans in case of changing circumstances.

Acoustic propagation studies have developed new insights into "ocean weather" effects on acoustic travel time, which will be important to the processing of ATOC data for climate trend direction. Coupled ocean climate models from the Princeton, Hamburg, and Massachusetts Institute of Technology research groups are being integrated with ATOC to provide better insights into the expected scales and distribution of global ocean change. ATOC is being funded by the Advanced Research Projects Agency (ARPA) from the Strategic

Environmental Research and Development Projects (SERDP) through a grant to Scripps Institution of Oceanography.

Box 9—Major ATOC Accomplishments

(1) Establishment of specific ATOC network configuration.

(2) Development and installation of hardware including survey of existing sites and acoustic characterization of network paths.

(3) Establishment of vertical line array designs.

(4) Agreement on an international cooperation plan.

(5) Scheduling of ocean "ground truth" measurement cruises for 1994 and 1995.

A Global Ocean-Atmosphere-Land System for Seasonal-to-Interannual Climate Prediction Program

A Global Ocean-Atmosphere-Land System (GOALS) for Seasonal-to-Interannual Climate Prediction Program was conceived as the U.S. contribution to the Climate Variability and Predictability (CLIVAR). CLIVAR is the WCRP's major new seasonal-to-interannual focused initiative. Planning for GOALS is being overseen by the National Research Council Climate Research Committee (CRC). The CRC has already completed a number of important steps, including producing a scientific background document and holding a major national scientific meeting with international representation. The CRC intends to establish a GOALS advisory panel in 1994, and GOALS will proceed with implementation plans during that year.

The ultimate scientific objectives of the GOALS program are to (1) understand global climate change variability on seasonal-to-interannual time scales; (2) to determine the extent to which this variability is predictable over time and space; and (3) to develop the observational, theoretical, and computational means to predict this variability, if feasible. A skillful forecast of average temperature and precipitation, a season to a year in advance, has already proven valuable in the countries in and around the tropical Pacific; an extension of predictions to the industrialized countries at higher latitudes would be of enormous economic benefit for agricultural planning, resource allocation, price support policies, and flood and drought mitigation.

The central hypothesis of GOALS is that variations in the forcing characteristics of sea-surface temperature, soil moisture, sea ice, and snow at the global boundary exert a significant influence on the seasonal-to-interannual variability of atmospheric circulation. Therefore, understanding variability and predicting climate at seasonal-to-interannual time scales requires understanding the processes that control these boundary conditions. Predicting the evolution of these boundary conditions will undoubtedly require improved models and observations.

The first phase of GOALS will augment the original prediction goals of TOGA by improving coupled models and by incorporating the data produced by the TOGA Observing System, especially the TAO Array, into predictions. Expansion into the global tropics will be based on the hypothesis that seasonal-to-interannual variability is related to variations in the locations, interactions, and effects of the major thermal sources and sinks. Expansion to higher latitudes will be guided by insights gained in studies of seasonal-to-interannual variability in the extratropical atmosphere, upper ocean, and land surfaces.

Following the successful example of the TOGA program, GOALS will be composed of four major program elements: modeling, observations, empirical studies, and process studies. The process studies will concentrate on the monsoonal forcing of the atmosphere in the eastern Pacific and Indian oceans and the transmission of these signals to higher latitudes. It is anticipated that two major ongoing TOGA activities, the TOGA Observing System and the TOGA Program on Seasonal to Interannual Prediction (T-POP) will be maintained during transition to the GOALS program. As GOALS matures, its programs can be expected to evolve and expand.

The success of GOALS will be measured in several ways: by the enhanced understanding of global climate variability and predictability on seasonal-to-interannual time scales, by the effectiveness of the observing system developed for describing and predicting the climate system, by the increased ability to model the processes involved in seasonal-to-interannual variations, and by the skill developed in predicting these variations.

Land-Ocean Interactions in the Coastal Zone

Coastal areas of the world are zones of increasing competition between the needs of human populations and the limited resilience of natural ecosystems. Human activities exert a tremendous burden on the coastal zone on both a global scale (e.g., sea-level rise and climate change) and a local scale (e.g., land use practices and overfishing). Because existing global change research lacks strong components focused specifically on the coastal zone, the International Geosphere-Biosphere Program developed the Land-Ocean Interactions in the Coastal Zone (LOICZ) program. LOICZ defines the coastal zone as extending from the coastal plains to the edge of the continental shelf. LOICZ is based on the premise that the current use of the coastal zone will inevitably affect its use by future generations. The creation of long-term, sustainable policies for coastal management will require an understanding of the many impacts derived from changes in climate, sea level, and land use and in the functioning of the ecosystems themselves. A LOICZ science plan has been published, and the implementation plan is scheduled for publication in late 1994.

The goals of LOICZ are: (1) to determine the fluxes of materials between land, sea, and atmosphere in the coastal zone, the capacity of coastal systems to transform and store particulate and dissolved matter, and the effects of changes in external forcing conditions on the structure and functioning of coastal ecosystems on global and regional scales; (2) to determine how changes in land use, climate, sea level, and human activities alter the fluxes and retention of particulate matter in the coastal zone and affect coastal morphodynamics; (3) to determine how changes in coastal systems, including responses to varying terrestrial and oceanic inputs of organic matter and nutrients, will affect the global carbon cycle and the trace gas composition of the atmosphere; and (4) to assess how the responses of coastal systems to global change will affect human use and habitation of coastal areas and to develop further the scientific and socioeconomic bases for the integrated management of the coastal environment.

U.S. involvement in the LOICZ program to date has been informal. Three U.S. scientists are members of the Scientific Steering Committee, but there is no official federal agency representation or program office in the United States. Efforts continue to determine appropriate U.S. participation.

The Geological Perspective

To understand the ocean system, it is necessary to describe and quantify the interconnections among its biology, chemistry, geology, and physics. Such an understanding does not currently exist; however, recent technological advances and ongoing observational and process-based research projects are beginning to provide detailed knowledge of oceanic conditions and processes. The geological record of oceanic and atmospheric conditions provides crucial information for constructing predictive models and interpreting data collected by the programs described in the previous section. The geological record cannot be understood, however, without a solid understanding of the processes by which ocean sediments and crust are changed by contact with seawater, including biological, chemical, and physical processes that transform dissolved and particulate material in the ocean before and after burial in the seafloor.

Even with a complete understanding of today's ocean, current climate models and our ability to predict future climatic changes would still be limited. To understand past, present, and future climate change fully, we must understand the long-term, natural variability of the ocean-climate system; the sensitivity of the ocean/climate system to change; and the reliability of models to predict future climate change. The geological record, in particular the paleoceanographic record, which details the development of the ocean system over time, is the sole source of much of the information needed to answer these questions.

Past oceanic conditions are discernable from the paleoceanographic record, deep-sea sediments, ice cores, shallow deposits from lakes and bays, and corals. Methods have been developed to collect and analyze data from these sources and to use these data to determine ocean and climate system change. Specifically, scientists attempt to reconstruct past temperature conditions in the ocean and atmosphere, chemical composition and circulation patterns of the ocean and atmosphere, changes in sea level, precipitation and evaporation, and biological productivity. These data provide a historical record of past environmental conditions over a range of time scales, putting present oceanic conditions into perspective.

There are several potential causes and effects of climate changes. Understanding the potential causes of climate fluctuations, including astronomic forces (e.g. the Milankovich cycle caused by Earth orbit variations), interactions within the ocean-atmosphere system, or tectonism and the opening and closing of ocean basins and connecting waterways, is essential for climate prediction. The geological record provides high-resolution data that may explain the processes acting to control and change the ocean-climate system. For example, past sea-level changes are attributed to several causes. However, geological data suggest that the rise and fall was primarily due to the modification of the ocean basins by tectonic activity. Additional cause-and-effect relationships need to be investigated, including the influence of the mid-ocean ridge system on ocean chemistry and circulation, the contributions of minerals on and in the seafloor (e.g. methane hydrates) to the ocean-climate system, and the relationship between atmospheric temperature and sea-level changes. This knowledge may allow scientists to quantify the effect that anthropogenic influences may have on climate and to determine whether human alterations can be controlled or reversed.

Continued progress in computer technology has improved scientists' ability to analyze data and to construct models of ocean-atmosphere and global climate systems. For example, the 1991-92 El Niño event, which brought a devastating drought to southern Africa and severe rains to southern California, Texas, and other parts of the world, was accurately predicted nearly 2 years in advance. Unfortunately, many models cannot predict events with such accuracy. The paleoceanographic record provides the unique opportunity to test the reliability of predictive mathematical models. The ability of models intended to describe the modern ocean to "hindcast" past ocean conditions accurately will provide confidence in their capacity to predict future oceanographic and climatic conditions. Hindcasting has already been used to test early atmospheric models. Paleoceanographers and modelers of the Climate Long Range Investigation Mapping and Planning (CLIMAP) program and Cooperative Holocene Mapping Project (COHMAP) tested the ability of atmospheric models to simulate past climates by utilizing data on predicted conditions from the glacial period that occurred 18,000 years ago. Similar tests can be performed on coupled ocean-atmosphere system models, which often lag in development behind atmospheric models. The reliability of climate predictions will almost certainly play a significant role in global environmental policy. The four programs described below focus on research programs designed to understand past climatic and oceanic conditions and the mechanisms that drive changes in these conditions.

Ocean Drilling Program

How good are climate predictions? This question can only be answered by checking predictions against the actual course of events. To understand the full spectrum of possible responses of the Earth's systems to disturbances from human activities, the geological record must be consulted. The most comprehensive long-term record of global climatic and oceanic change comes from the sediments of the world's ocean. The Earth's paleoceanography is reconstructed through the analysis of deep-sea cores collected from around the world by the Ocean Drilling Program (ODP). Information extracted from ODP cores makes it possible to reconstruct past climatic and oceanic conditions, and, more importantly, to begin to understand the mechanisms that drive these changes. A few recent accomplishments of the program are listed in Box 10.

The Ocean Drilling Program is an international program that uses the products from the 470-foot drilling vessel JOIDES *Resolution* to answer questions about past and present processes of the Earth. Drilling is based on suggestions and proposals from the entire scientific community. The program provides core samples and data from downhole experiments in the ocean basins, as well as facilities for the study of these samples and data—on the ship and on the shore.

The deep-sea record contains the critical evidence for large-amplitude, short-term climate change. The study of cores from different oceanographic regimes related to physical, chemical, and biological gradients across latitudes and depth is the most efficient method to capture the response of the ocean's systems to global climatic change. Since 1988, one of the ODP's major operational goals has been to obtain continuous sedimentary sections from key places in the world oceans. Partial depth transects have been collected in the western and northern Pacific Ocean, the Tyrrhenian Sea, the eastern equatorial Atlantic, the Peru Margin, the Maud Rise, the Subantarctic, the Indian Ocean, the tropical Pacific, the Santa Barbara Basin, and soon the tropical Atlantic, the subarctic Atlantic, and the Norwegian Sea. Other proposed targets include the California Current, the Walvis Ridge, the western equatorial Atlantic, and the high Arctic.

As a result of these studies, it is known that biotic, geochemical, and sedimentary cycles that correlate with periodic changes in the Earth's orbit are pervasive in the marine record of past climates. Orbital signatures may vary spatially and evolve through time, but they have major implications for mechanisms of climate change. The Neogene (past 24 million years) is the period when present-day oceanic conditions became established, and, based on ODP studies, conceptual frameworks have begun to emerge to explain Neogene climate change. However, the details of competing hypotheses—and their global implications—remain to be tested. Testing competing theories will require recovery of sediments over a wider range of geography and depth; sediments accumulating at very high rates, such as on continental margins and in enclosed ocean basins; and sediments from high latitudes.

The ODP uses deep-sea cores to study how the oceans respond to Earth's changing climate. Key foci include: (1) understanding how changes in ocean circulation, ocean chemistry, and biological fluxes influenced climate and atmospheric pCO_2 during the Neogene; (2) determining causes of anomalously warm global climates; (3) detecting changes in hemispheric thermal gradients and circulation intensity; (4) determining causes of the initiation of continental glaciation; and (5) discovering the mechanisms for forming carbon-rich and/or anoxic sediments. Results of ODP studies have raised serious questions that challenge modern climate theory. For example, the ODP has recovered convincing evidence for the existence of warm salty bottom water sources in the ancient oceans—this poses new questions about sources, distribution, and mechanisms of watermass formation that remain unanswered and will require additional study.

During the past 5 years, oceanographers have become more aware of the critical role of biological productivity of the oceans in influencing global climate. Drilling in high-production areas such as the Peru and Oman margins, the subantarctic Atlantic, northwest Africa, and the equatorial Pacific and Atlantic has greatly enlarged the distribution of samples suitable for productivity studies. Because of ODP advances in analyzing past oceanic productivity through the recovery of complete sedimentary sections, it is now possible to express sedimentary data in terms of actual fluxes through time. Concurrently, under the auspices of other oceanographic programs such as JGOFS, the role of biological productivity on ocean chemistry and climate is becoming better understood. In combination, these advances have led to significant improvements in our ability to model paleoceanographic and paleoclimatic conditions.

The ODP is interested in improving the understanding of global sea-level history and the causes of sea-level change, since few aspects of global change are unrelated to sea-level change. However, knowledge about global sea-level change is still primitive. It has become clear that glacially-induced sea-level changes of hundreds of meters have occurred during the past 30 million years, but sea-level changes recorded in rocks during the preceding 500 million years have not been sufficiently studied. Clearly, we must understand the mechanisms of sea-level change before we will be able to predict the course of sea-level change over the next few decades, or even over millennia. The ODP has made significant progress toward understanding the mechanisms and timing of global sea-level change by implementing focused drilling programs on mid-oceanic islands (atolls) and continental margins.

The scientific community interested in ODP's contribution to global change studies is growing rapidly; liaisons are being established between the ODP and other geological global change programs such as the U.S. Marine Earth System History (MESH) program and its international counterpart MAGES, which is associated with PANASH (Paleoclimate of the Northern and Southern Hemispheres) of the IGBP Past Global Climate (PAGES) program. All of these programs focus on the causes and consequences of global climate change during the past 500,000 years, but MESH also focuses on the warm climates of the Pliocene and the early Eocene. To study the long-term evolution of climate and

Box 10—Recent ODP Accomplishments

(1) ODP leg 151 acquired over 3 km of core in the Norwegian-Greenland sea and the Arctic Ocean in the summer of 1993 to provide information on the glaciation of Greenland and on the influence of the Arctic Ocean on global climate change.

(2) At four sites on the eastern continental slope of the United States, leg 150 dated numerous geologic boundaries that were caused by sea-level changes. These sites are part of a transect of holes that will extend in the future across the shelf. This transect will test models describing how the margin responds to sea-level changes.

(3) Legs 149 and 152 were drilled off Iberia and Greenland to examine the influence of passive rifting (Iberia) and volcanism (Greenland) on the opening of the North Atlantic Ocean.

(4) On the Cascadia margin of the U.S. West Coast, leg 146 drilled a series of shallow holes to examine fluid flow through the margin and the relationship between this flow and gas in the sediments. The boundary between the solid (ice) and gas phases was penetrated and measurements made in situ.

(5) Leg 145 documented an onset of volcanism 2.6 million years ago, which also appeared to mark the onset of glaciation in the northern hemisphere.

(6) In moves designed to internationalize the ODP to a greater degree, a new core repository was established at the University of Bremen in Germany, and the JOIDES office will move to the United Kingdom in 1995.

environmental variability under different climatic conditions will undoubtedly require continued scientific drilling by the ODP. These studies will lead to a better understanding of the conditions in the ancient oceans and of climatic changes through time and, in turn, to a fuller comprehension of the continuing evolution of Earth's environment.

Funding for ODP is provided by NSF together with contributions from international partners including the Canada/Australia Consortium, the European Science Foundation, the Federal Republic of Germany, France, Japan, and the United Kingdom. Joint Oceanographic Institutions, Inc. (JOI) is the prime contractor. JOI subcontracts to Texas A&M University, which leases, operates, and staffs the drillship JOIDES *Resolution* and maintains facilities for storage and study of cores. Lamont-Doherty Earth Observatory of Columbia University is the contractor responsible for downhole logging measurements.

Summary

The sediments at the bottom of the ocean record in detail and over a long period of time (more than 100 million years) the history of our planet. This record is dominated by climate changes, including those induced by changes in ocean circulation, glaciation, impacts of celestial bodies (asteroids and comets) on Earth, volcanic eruptions, and plate tectonics. ODP is essential to deriving Earth history information from the deep-sea sediments and rocks. Without the capacity to drill and sample in the deep ocean, this important branch of Earth Science would not exist.

Ridge Inter-Disciplinary Global Experiments

The Ridge Inter-Disciplinary Global Experiments (RIDGE) initiative is unique among the many programs studying the role of the ocean in global change because it addresses the volcanic and tectonic cycles, which are driven by the terrestrial energy source rather than the sun. Significant amounts of heat and chemicals are exchanged between the ocean and the solid Earth via the global mid-ocean ridge system. Much of this exchange is accomplished through hydrothermal activity, which results from the circulation of seawater through the oceanic crust. During this process, seawater changes composition as it circulates through, and reacts with, the young lavas. As it heats up, it becomes buoyant, resulting in the emanation of hot, acidic fluids on the seafloor. These hot springs are associated with very unusual, diverse biological communities that are sustained by chemosynthesis. The fluxes associated with this activity have an important, but as yet unquantified, impact on ocean circulation, heat transport, seawater chemistry, ocean ecosystems, and ocean-atmosphere interactions, all of which may in turn affect global climate.

RIDGE is designed to understand the causes and to predict the consequences of mantle-driven physical, chemical, and biological fluxes within the global spreading center system. Through integrated observational, experimental, and theoretical studies, it seeks to determine the primary processes that have shaped the physical evolution of our planet and to describe long-term temporal variations that may have modified the past climate of the Earth. For example, geological evidence indicates that early in Earth's history there was a cooler sun but high atmospheric carbon dioxide concentrations were critical in keeping the Earth warm. It is therefore of fundamental importance to understand the rate of release of carbon dioxide during mid-ocean ridge volcanism, and the factors controlling its variability, in order to determine its role in the global carbon cycle and in climate change.

On shorter time scales, the distribution of hydrothermal activity along the mid-ocean ridge system, its episodicity and variability, play important roles in regulating ocean chemistry and circulation—two of the critical factors in predicting global climate change. Within RIDGE, there are several research initiatives that are investigating these factors through reconnaissance surveys of previously unexplored parts of the global ridge system, development of response strategies to the detection of events on mid-ocean ridges, and detailed measurements of temporal variations at a single "observatory" site.

Given the significant exchange of heat and mass between the solid Earth and the ocean, studies of the mid-ocean ridges are an integral part of the study of the ocean-atmosphere system and of global change. The RIDGE Initiative has been formally identified as an element of the U.S. Global Change Research Program, with funding from NSF. In addition, NOAA, the Navy, and the U.S. Geological Survey (USGS) have programs focusing on the dynamic processes along mid-ocean ridges. The RIDGE Initiative has recently been expanded internationally (InterRIDGE), and several countries have developed their own equivalent national research efforts. InterRIDGE has a working group sponsored by SCOR.

Since 1989, when the Initial Science Plan for the RIDGE program was published, progress has been made in several of the research areas identified as initial foci of the program (Box 11). One long-term goal of the program is to obtain sufficiently detailed spatial and temporal definitions of the global mid-ocean ridge system to construct quantitative, testable models of how this system works. These models could then be used to predict the impact of variability in ridge processes on global climate change. An essential element of this strategy is the identification of key variables that affect the crustal accretion process. The RIDGE program has focused initially on the interplay of spreading rate and magma supply as two first-order variables. With the goal of collecting two comprehensive, comparable datasets—one in a fast-spreading and one in a slow-spreading regime—field programs began in 1991 to collect data on the slow-moving Mid-Atlantic Ridge. During Phase I of the French-American Ridge Atlantic (FARA) project (1991-93), the purpose of the studies was to define the overall architecture of the plate boundary and to describe its first-order tectonic, bathymetric, petrologic and hydrothermal characteristics. This program has been successful and has resulted in virtually continuous and routine geophysical coverage of the axial zone between 15°N and 40°N, together with densely spaced rock samples for geochemical studies. In addition, cruises have prospected for hydrothermal plumes using towed vehicles carrying physical and chemical sensors and combined conductivity-temperature-depth (CTD) and rosette sampling. This effort has led to the discovery of two new hydrothermal sites, one on shallow crust near the Azores, and the other just south of the Atlantis Fracture Zone, thereby doubling the number of known active hydrothermal sites on the Mid-Atlantic Ridge. During Phase II of the FARA project (1993-95), more focused studies will concentrate either on smaller areas identified to be of particular interest, such as the new hydrothermal sites, or on specific experiments, such as detailed seismic investigations.

One of the major challenges for the RIDGE program is to develop practical and reliable methods for detecting, locating, and responding to transient ridge crest phenomena, such as volcanic eruptions, tectonic activity, or catastrophic releases of heat or chemical mass into the water column (Box 11). Monitoring of subaerial eruptions in Hawaii and Iceland has led to major advances in understanding how these systems work. Observation of an actively spreading portion of the mid-ocean ridge system will result in a better understanding of the sequence, duration, and type of activity and of the interrelationships among the volcanic, tectonic, hydrothermal, and biological processes involved in the creation of new crust.

An important accomplishment of RIDGE over the last few years has been the monitoring of a site on the East Pacific Rise crest where a very young volcanic eruption was serendipitously discovered in 1991 during submersible operations in the area. The eruption was associated with abundant and widespread venting of hydrothermal fluids. Mineral deposits and biological communities typically seen at such vents had not yet developed; however, white bacterial mats covered the fresh, glassy lava flows, and the overlying water was filled with white bacterial matter. A return to the site in 1992 found dramatic changes, with venting more focused at localized high-temperature vents, changes in the temperature and chemistry of the fluids, less bacteria, and a diverse biological community, including tubeworms up to 30 cm in length. This area will continue to be revisited over the next couple of years to observe the changes that occur as the lava cools and the hydrothermal system matures. However, it is clear from the rapid changes observed in the first year following the eruption that early detection and rapid response are critical in studying the first stages in the formation of the oceanic crust.

Progress is being made in early detection of ridge events. In 1991, NOAA's VENTS program began continuously recording hydrophone data from a network of permanent deep-ocean hydrophone array (SOSUS) operated by the U.S. Navy and located in the northeast Pacific. These sophisticated sensing devices are designed to monitor sound in the SOFAR (sound fixing and ranging) channel and are ideally distributed for monitoring and locating activity along the spreading centers off the coasts of Washington and Oregon. Initially, these data were recorded and sent to NOAA for identification and location of specific events. However, in June 1993, a system was installed that allowed real-time monitoring of the Juan de Fuca Ridge and, within 4 days, a burst of seismic activity was observed near Axial Seamount. A Canadian survey ship in the area confirmed the presence of hydrothermal plume signals. Since then, studies have continued with a short response cruise in late 1993. This new capability for ridge event detection will be available to the scientific community in the near future and will almost certainly lead to new and exciting discoveries in the North Atlantic.

Over the next few years, the RIDGE Initiative will see the implementation of several major field programs and experiments. Estimation of the energy and chemical fluxes associated with ridge processes is vital to assessing their impact on global climate change. However, these estimates require characterization of the tectonic structure, geochemistry, biology, and energy fluxes of the mid-ocean ridge on a global scale—a goal possible only in the context of a concerted international effort coordinated through InterRIDGE. The general approach will be to identify 1,000-2,000-km-long sections of the ridge that are located in the least well-characterized portions of the oceans. In the next 2 years, RIDGE field programs will be conducted along the Southeast Indian Ridge in the vicinity of the Australia-Antarctic Discordance as part of this international effort.

Another important, but poorly understood, aspect of the spreading process is melting beneath the ridge axis and the mechanism of melt migration through the oceanic lithosphere to the seafloor. Since 1989, RIDGE scientists have been designing a field experiment aimed at investigating the size and geometry of the melting region, mantle flow, and melt migration beneath the mid-ocean ridge. This project, the Mantle Electromagnetic and Tomography (MELT) experiment, will involve deployment of up to 40 ocean bottom seismometers to record the effects of near-ridge structure on signals from distant earthquakes and up to 20 electromagnetic instruments to determine the conductivity structure of the crust and uppermost mantle for time periods of up to a year. An experiment of this scale will require detailed coordination of many groups within the United States and utilization of almost the entire inventory of these specialized instruments. The location chosen for this study is the East Pacific Rise south of the Garrett Fracture Zone—one of the fastest spreading portions of the mid-ocean ridge system—and it is expected that sea-going operations will commence in 1995.

One important long-term goal of the RIDGE program is to understand interactions among ridge crest processes on time scales of seconds to decades. This requires time-series observations of magmatic, volcanic, tectonic, and hydrothermal phenomena at selected sites along the global ridge system. A start has been made with the response effort on the East Pacific Rise that was discussed above. However, a second, high priority approach of the RIDGE program is the initiation of a seafloor observatory using instrumentation capable of long-term monitoring of volcano-hydrothermal systems. The site selected for this effort is the Cleft Segment on the Juan de Fuca Ridge, and a number of instruments and moorings have already been deployed in this region to look at spatial and temporal variability in volcanic, tectonic, and hydrothermal processes. It is expected that, over the next few years, a coordinated effort will result in the simultaneous measurement and monitoring of a number of key variables (e.g., ground deformation, microearthquake activity, flow and chemical composition of vent fluids, distribution of biological organisms) that will allow estimation of energy and mass fluxes from an individual ridge segment over a short time scale. This will be an important step in determining the role of these systems in regulating ocean chemistry and circulation—two critical factors in predicting global climate change.

Many aspects of the research conducted within RIDGE provide fundamental knowledge necessary for the nation's technical and economic development. For example, study of submarine hydrothermal systems is leading to a new understanding of how ore deposits form and may lead to new methods of exploration. The thermophilic bacteria that inhabit submarine vents are providing enzymes that have spawned a new area of biotechnology allowing genetic chemistry to be conducted at boiling temperatures. Currently a $600 million per year industry, the market for these enzymes is expected to grow dramatically in the next decade. The seafloor mapping and deep submergence technology developed in ridge-related research is already a billion dollar business being widely used in marine-related industries ranging from petroleum exploration to fisheries.

Box 11—Recent RIDGE Accomplishments

(1) Development, in collaboration with the NOAA Vents Program, of a practical and reliable event detection and response capability using a land-based acoustic monitoring system (SOSUS) and in 1993 the first successful detection of a ridge crest volcanic eruption.

(2) Monitoring of a very young volcanic site on the northern East Pacific Rise that has documented, for the first time, the volcanic, tectonic, hydrothermal, and biological processes occurring in the immediate aftermath of a major seafloor eruption.

(3) Characterization, as part of the French-American Ridge Atlantic (FARA) project, of the Mid-Atlantic Ridge between 15°N and 40°N with continuous multibeam bathymetric mapping, densely spaced rock samples and geochemical studies. This work has doubled the number of known hydrothermal sites in the Atlantic.

(4) Completion of site surveys and pilot studies along the southern East Pacific Rise for the Mantle Electromagnetic Tomography (MELT) experiment. MELT is the most ambitious marine geophysical experiment ever attempted and is aimed at determining the size and geometry of the melting region beneath a ridge crest.

(5) Promotion of InterRIDGE, a coordinated, international program of ridge crest research that currently involves over a dozen countries.

Summary

The goal of the RIDGE program is to understand the geology, physics, chemistry, and biology of processes occurring along the global mid-ocean ridge system. The mid-ocean ridge is the largest continuous geological feature on the planet and it is the surface expression of convective processes occurring within the Earth's mantle. Volcanism along the mid-ocean ridge creates the oceanic crust, which forms 60 percent of the Earth's surface. These volcanic and related hydrothermal processes play a critical role in controlling the thermal evolution of the planet, regulating the chemistry of seawater, and providing the energy source to support a diverse and unique biological community. Progress has been made in several research areas including obtaining detailed spatial and temporal information of the ridge system and the detection and monitoring of ridge crest phenomena (Box 11).

The Global Change and Climate History Program of the U.S. Geological Survey

To address national and international concern over the prospect of environmental changes due to human activities and to provide relevant information on global climate change to the government and the research communities, the Geologic Division of the U.S. Geological Survey conducts research in paleoclimate and current climatic processes. The Global Change and Climate History Program focuses on five primary research elements: paleoclimate, cold regions, arid and semi-arid regions, biogeochemical dynamics, and volcano emissions.

The role of the ocean in global climate change is addressed primarily in three research areas: (1) Pliocene climate reconstructions, (2) Arctic paleoceanography, and (3) integration of marine and terrestrial records of climate change in western North America. Accomplishments and future plans in these research areas are presented in Box 12.

As part of a multidisciplinary study to map environmental conditions during the mid-Pliocene warm intervals (2.5-3.5 million years ago) and to document variability during the Pliocene, an effort is under way to derive geological proxy measures for oceanographic properties such as sea-surface temperature and deep-ocean circulation. The results from this study, combined with those from a coordinated effort studying vegetation in terrestrial sites, will be used to constrain and validate the results of general circulation model experiments to "hindcast" warm intervals of the Pliocene.

To document the history of the Arctic Ocean during the Late Neogene (past 5 million years), another multidisciplinary study is in progress. The study includes seismic profiling, stratigraphic and sedimentological observations, and detailed climate history research using micropaleontology and stable-isotope analysis of cores recovered from the Arctic basin.

To improve predictions of future climates, a study is under way to understand the climate history of the eastern North Pacific Ocean and the adjacent land areas of the western United States over the past 130,000 years. A primary objective of the study is to provide detailed correlations of marine and terrestrial climate records along a transect extending from the western interior of the United States to the eastern Pacific Ocean. A full range of paleontological, sedimentological, geochemical, isotopic, and chronological studies and techniques are being used to reconstruct the climate history of the region.

The results from these specialized studies will add to a more complete understanding of the world ocean and of the potential changes that it may undergo in the future. The USGS effort is integrated with other federal agencies through the U.S. Global Change Research Program, coordinated by the Committee for Earth and Environmental Sciences. Studies in the Arctic involve formal cooperation with the Geological Survey of Canada.

Box 12—USGS Program Accomplishments and Future Plans

(1) Pliocene climate reconstruction: A preliminary set of mid-Pliocene northern hemisphere boundary conditions (sea-surface temperature and vegetation) have been completed and used in a simulation of Pliocene warm climates with the GISS global circulation model. The model simulations indicate that a 15-20 percent change in meridional ocean heat transport (from low to high latitudes) may have been an important component of Pliocene warming. The boundary condition data sets are being refined, updated, and extended into the southern hemisphere. Additional modeling and sensitivity experiments incorporating the revised paleoenvironmental data sets and at least two different models are planned for 1994.

(2) Arctic paleoceanography: In 1992 a USGS-sponsored geophysical and coring expedition with the USGS Icebreaker *Polar Star* collected 52 piston cores and 17 box cores from the North Wind Ridge and adjacent areas of the Canada Basin in the western Arctic Ocean. Initial analyses of the 1992 cores, and of cores collected in the same area by the USGS in 1988, indicate that the sediments of the North Wind Ridge contain a record of glacial-interglacial cycles for the last few million years. A combination of lithostratigraphy, magnetostratigraphy, and quantitative microfossil analyses allows construction of a composite section containing a climatic and paleoceanographic record for the western Arctic that extends back into the Pliocene. Future plans include detailed analyses of existing cores and additional sampling.

(3) Marine terrestrial records of western North America: In 1992 the USGS began a coring program for climate records in the western United States to obtain terrestrial climate records that could be compared with climate records for marine cores of the eastern North Pacific. In the last 2 years cores of lacustrine deposits have been recovered from the Klamath Lakes region of southern Oregon, Butte Valley in northern California, Owens Lake in California, and southern Idaho. Future drilling targets include the Bonniville Basin and the Yukon Basin in Alaska. The climate records developed from these cores will be correlated and compared to climate records from marine cores from the eastern North Pacific and Arctic oceans.

Marine Aspects of Earth System History

The objective of the Marine Aspects of Earth System History (MESH) program is to understand long-term natural variation in global environments, which are recorded in the ocean's geologic record. The 10-year program focuses on the dynamics of the coupled ocean-climate system, including sensitivity to change over a range of time scales, responses to external forcing, and internal variability. Through generation and analysis of data on the Earth's response to various types of forcing (both internal and external to the climate system) and experimentation with climate models, critical environmental feedback mechanisms, such as the role of ocean chemistry and the greenhouse effect, will be better quantified. This will yield improved understanding of global change processes, and better models for predicting future environmental changes on the scale of tens to thousands of years. This information in turn will be important to policy discussions about global change, including society's attempts to adapt to, or mitigate the effects of, future change.

The initial MESH Advisory Panel was convened in 1990 and proposed five broad programmatic themes that emphasized the marine geological aspects of global change. The report from this meeting served as the springboard for the panel to solicit white paper proposals from the ocean sciences community to address how the marine geologic record could be used to help understand the process and record of global change. At a workshop held in March 1993, the white papers were discussed, scientific priorities were defined, and a MESH Steering Committee was elected. The Steering Committee met in September 1993 to refine the results of the March meeting and to develop a MESH Program Plan.

The MESH Advisory Panel and Steering Committee identified and prioritized eight themes: (1) ocean geochemical dynamics and climate change, (2) climate sensitivity and variability at time scales of 10^3 to 10^5 years, (3) extreme warmth, (4) marine records of seasonal to millennial scale variability, (5) abrupt climate change, (6) polar cryospheric history and global climatic change, (7) marine records of continental climate change, and (8) biological response to climate change. In ranking the eight themes, it was recognized that significant overlap exists between the scientific objectives of individual themes. For example, knowledge of the state of the polar cryosphere is an essential part of understanding times of extreme warmth. Taking these overlapping priorities into account, MESH identified five program elements that address the highest priorities of the paleoceanographic community. These highest priority research themes and their objectives are:

(1) Ocean geochemical and climate change. To better understand how ocean circulation, ocean chemistry, and biologic fluxes have influenced the observed record of atmospheric CO_2 over the past 500 thousand years and the consequences of these changes to the ocean, atmosphere, and biosphere system.

(2) Climate sensitivity and variability at time scales of 10^3 to 10^5 years. To study the response of the ocean to a known, strong, direct forcing (changes in the distribution of solar radiation and atmospheric CO_2 content) in order to identify processes that control variability in ocean circulation and chemistry; and to use paleoceanographic histories along with climate models to identify the sensitivity and character (i.e., quasi-linear to nonlinear internal feedbacks) of these climate related processes.

(3) Extreme warmth. To better characterize episodes of extreme climatic conditions over the past 120 million years, including features of the ocean and atmospheric chemistry, with a focus on a selected set of short, stable intervals of warm climates; and to document changes in the climate system and develop models of these climate episodes in order to understand the origin of these extremes, identify the feedbacks within the climate system that maintained global climate in the extreme states, and articulate the mechanisms that brought an end to these extreme conditions.

(4) Marine records of seasonal to millennial scale variability. To understand how the climate system operates on societal (seasonal to millennial) time scales, including the response of climate variability to past changes in boundary conditions; and to assess the abilities of predictive numerical models to simulate seasonal to millennial climate variability and sensitivity.

(5) Abrupt climate change. To identify and characterize isolated events and transitions in the geologic record that can help in evaluating the processes and feedbacks that determine the degree of sensitivity and stability of the Earth system.

A Program Plan is being submitted by the Steering Committee for MESH. It represents the first step in implementing the scientific priorities defined in the MESH report, the solicited white papers from the community, and the discussions and report of the March 1993 meeting. MESH is partly organized under the scope of the International Geosphere-Biosphere Program's Past Global Climate (IGBP PAGES) program and is the U.S. component of the international program IGBP International Marine Global Changes Study (IMAGES). MESH will be funded as part of the NSF Ocean Sciences Program.

Summary

The objective of MESH is to understand long-term natural variations in global environments, which are recorded in the ocean's geologic record. The 10-year program focuses on dynamics of the coupled ocean-climate system, including sensitivity to change over a range of time scales, responses to external forcing, and internal variability. Through generation and analysis of data and experimentation with climate models, critical environmental feedback mechanisms such as the role of ocean chemistry and the greenhouse effect will be better quantified. This will yield improved understanding of global change processes, and thus better models to predict future environmental changes on the scale of decades to millennia. This information will be useful to policy discussions for global change, including society's attempts to adapt to future changes, or to mitigate its effects.

Crosscutting Issues

Platforms for Observation and Collection

Attainment of the challenging goals of the programs described in this report depends on the collection and analysis of materials and data. A wide variety of techniques, equipment, and research vessels are required for this task. The availability of satellites and in situ moorings will not diminish the importance of surface vessels in oceanographic research. Ships will be necessary for collection of subsurface samples over wide areas, as adjuncts to satellites (for "sea truthing") and to long-term moorings and drifters (for placement, retrieval, calibration, and validation studies), and for instrument development. For example, the collection of core material and downhole measurements requires sophisticated equipment and an advanced drilling vessel. ODP uses the JOIDES *Resolution*, which was modified to meet the special requirements of scientific ocean drilling and replaced the drillship *Glomar Challenger* in 1985. The *Resolution* has the capability to drill deeper, in more difficult rock formations, and with a more comprehensive set of logging experiments than did the *Challenger*.

Since the mid-ocean ridges are often located under more than a mile of water, deep-sea research vessels such as the *Alvin* submersible are required for sampling. The *Alvin*, which was involved in the discovery of deep-sea hydrothermal vents near the Galápagos Islands in 1977, has made hundreds of dives to vent sites along the global ridge system providing scientists with a unique opportunity to investigate these environments. Remotely operated vehicles (ROVs) and autonomous underwater vehicles (AUVs) are being used with increasing frequency for undersea observations and sampling. Improved optics, increased power of integrated circuits, new materials, and new sensors have made ROVs and AUVs important tools for oceanography. As new sensors become available and size and power requirements decrease, these unmanned vehicles could provide even greater capabilities. The primary limitation at present is the high cost of sensors.

Natural laboratory sites for long-term, in-depth studies of the seafloor have been established by the Office of Naval Research on the East Pacific Rise and the Mid-Atlantic Ridge. ONR's natural laboratories are located on spreading centers along the mid-ocean ridge system and will yield valuable geological and geophysical data. The value of natural laboratories will increase as the variety of experiments and surveys increases.

Long Time Series

It has been demonstrated that continuous or regular observations of ocean conditions at specific sites must be collected over long time intervals (years to decades) in order to characterize the natural variability of oceanic conditions. Ships cannot provide this kind of monitoring. Long-time-series data for certain ocean surface parameters (wind, temperature, color, sea-ice cover) can be collected by satellite sensors, while additional parameters can be measured by in situ sensors. Satellite sensors can observe a specific site with overflights on the order of a few times per week, observations may be limited by cloud cover, and typical sensors last only a few years. It has been difficult to maintain data continuity because of the long lead times for sensor and satellite development and the vulnerability of spacecraft to the federal budget process (satellites are discussed in more detail in the next section).

In situ instrument moorings obviously provide less global coverage than satellites, but can measure conditions repeatedly at one site from the surface to the bottom for months at a time. The major constraints to using in situ instruments for long-time-series observations are the harshness of environmental conditions, limited availability of power, and lack of appropriate sensors for measuring some parameters. Significant advances in sensor development are expected in the next 15 years, but for now the lack of in situ instrumentation is a serious limitation.

Many of the required long-time-series observations will be obtained by the Global Ocean Observing System (GOOS), which is specifically intended to provide ongoing observations of critical variables. Some specialized measurements initiated, developed, and operated by other research programs will eventually be transferred to GOOS (e.g., the TOGA observing systems). Just as the World Weather Watch has provided a background series of observations to enable global-scale meteorological research, it is expected that GOOS will provide a background of long-time-series observations for oceanographic and climate-change research. GOOS will also be a vehicle for the development of some of the required instrumentation.

Satellites

Satellites are an extremely important component of many of the programs described in this report because they provide the only opportunity to observe much of the ocean surface in periods of less than a day. It is impossible to achieve synoptic coverage from ships. However, field measurements must still be carried out at selected locations to correlate remote satellite observations with conditions on and beneath the ocean surface as measured from ships and other platforms. Several of the programs described in this report have been designed to exploit particular satellite sensors for monitoring large-scale conditions occurring coincident with shipboard measurements.

JGOFS will incorporate ocean color image data from the Sea-Viewing, Wide Field Sensor (SeaWiFS), and scatterometer data from the Advanced Earth Observing Satellite (ADEOS) and European Remote Sensing Satellite (ERS-1). Ocean color is related to phytoplankton biomass, whereas the scatterometer measures the scattering of radiation by the ocean surface, which is related to waves, and in turn to wind speed. Estimates of wind and waves are necessary to estimate heat, momentum, and gas (i.e., CO_2) transfers across the ocean surface. Wind stress is the primary driving force for the upper ocean circulation. Thus, scatterometer wind observations are needed to test predictive models of ocean circulation and to develop improved data for use in climate models. Sea-surface temperature can be measured by infrared radiometers (for example, on ERS-1) and combined with wind measurements to estimate ocean-atmosphere heat exchange.

The TOGA program and WOCE require satellite measurements of sea-surface topography and winds. The Geodetic Satellite Mission (GEOSAT) demonstrated the ability of altimeters to observe the tropical ocean waves generated by El Niño/Southern Oscillation (ENSO) events and to show changes in surface currents driven by the slope of the sea surface. The TOPEX/Poseidon altimeter is providing sea-level measurements three times more accurate than those from GEOSAT or ERS-1, although ERS-1 will make measurements to somewhat higher latitudes than TOPEX/Poseidon. The accuracy of the TOPEX/Poseidon altimeter will allow it to play a central role in measuring surface currents and lateral oceanic heat transport. The NASA scatterometer (NSCAT) that will be included on the Advanced Earth Observing Satellite (ADEOS) will be more accurate and will provide data for a greater percentage of the ocean surface than will the ERS-1 scatterometer.

NASA is developing a series of large and small satellite platforms on which to fly ocean sensors, the Earth Observing System (EOS) and Earth Probes. These sensors will measure clouds, rainfall, surface temperature, atmospheric particles, and snow and ice cover. The first Earth Probe satellite will support the SeaWiFS ocean color sensor, followed by the Tropical Rainfall Measuring Mission, which will provide data to estimate heat transfer by evaporation and precipitation of water in the tropics.

Data Management and Availability

GOOS and GCOS will cooperate to develop the next generation of atmosphere-ocean data management systems. The Global Telecommunications System (GTS) was developed for the World Weather Watch and is now showing its age. The new system will be Internet-based and will involve distributed data bases. One significant aspect of the trend toward GOOS and GCOS as major observational programs is their provision of data for the public good. Since most of the data will have real-time or near real-time uses, as well as retrospective purposes in research programs, it is essential that the data be available in real-time, with no restrictions on dissemination and no proprietary periods during which the data are held by a single researcher. The guiding principle is that all data should be, to the extent technically possible, available immediately for use by all. This principle dictates an increasing use of satellite telemetry systems and advanced processing and quality control systems.

Conclusions

Major advances in scientific understanding are often enabled by new technology that provides a clearer view and different perspective of natural processes. When these advances have an impact on societal and economic issues, major national benefits can accrue. Today, the field of oceanography has an opportunity to assemble new data and impart new understanding of the ocean's role in global change. Many aspects of our society may depend on how well we can anticipate the role that the ocean will play in determining the future climate of the planet. Our ability to anticipate, and possibly avoid, deleterious changes in global climate will depend on our ability to understand how the biological, chemical, physical, and geological processes in the ocean interact with each other and with atmospheric and terrestrial processes. This depth of understanding will be possible only by implementing the carefully planned programs devised by oceanographers in academia and government and by providing access to the necessary "tools"—ships, satellite sensors, in situ instruments, oceanic cores, and computer resources—to successfully complete these programs.

The Ocean Studies Board (OSB) published a report in 1992, *Oceanography in the Next Decade: Building New Partnerships*, that made a number of recommendations related to the programs described herein. The OSB recommended that "the academic institutions, individually or through consortia, take a greater responsibility for the health of the field including nationally important programs." In particular, the large, long-lived global change research programs highlight the need for institutional responses that are more stable and of longer duration than those of individual scientists. The OSB also recommended that "academia and federal agencies work together to ensure that appropriate long-term measurements are extended beyond the work of any individual scientist or group of scientists and that the quality of such measurements is maintained."

A partnership among academia, government agencies, the private sector, and large-scale international and national global and climate change programs should be established. Since the primary goal of global change programs is accurate climate prediction and since the future of our society may depend upon achievement of this goal, global change programs should be carefully planned, thoroughly reviewed, and sufficiently funded.

APPENDIX I - List of Acronyms

ACCP	Atlantic Climate Change Program
ACP	Antarctic Circumpolar Current
ADCP	Acoustic Doppler Current Profiler
ADEOS	Advanced Earth Observing Satellite
ALACE	Autonomous Lagrangian Circulation Explorer
ARCSS	Arctic Systems Science
ARPA	Advanced Research Projects Agency
ATOC	Acoustic Thermometry of Ocean Climate
AUV	Autonomous Underwater Vehicle
CFC	Chlorofluorocarbon
CLIVAR	Climate Variability and Predictability
CRC	NRC Climate Research Committee
CTD	Conductivity-Temperature-Depth
CZCS	Coastal Zone Color Scanner
DOE	Department of Energy
ENSO	El Niño-Southern Oscillation
EOS	Earth Observing System
ERS-1	European Remote Sensing Satellite
FARA	French-American Ridge Atlantic
GCOS	Global Climate Observing System
GCP	GLOBEC Core Program
GEOSAT	Geodetic Satellite Mission
GFDL	Geophysical Fluid Dynamics Laboratory
GISP2	Greenland Ice Sheet Project Two
GLOBEC	Global Ocean Ecosystem Dynamics
GOALS	Global Ocean-Atmosphere-Land System
GOOS	Global Ocean Observing System
GTS	Global Telecommunications System
ICES	International Council for the Exploration of the Sea
ICSU	International Council of Scientific Unions
IGBP	International Geosphere-Biosphere Program
IMAGES	International Marine Global Changes Study
InterRIDGE	International RIDGE
IOC	Intergovernmental Oceanographic Commission

IRICP	International Research Institute for Climate Prediction
JGOFS	Joint Global Ocean Flux Study
JOI	Joint Oceanographic Institutions, Inc.
LAII	Land-Atmosphere-Ice Interactions
LOICZ	Land-Ocean Interactions in the Coastal Zone
MELT	Mantle Electromagnetic and Tomography
MESH	Marine Aspects of Earth System History
NASA	National Aeronautics and Space Administration
NSCAT	NASA Scatterometer
NOAA	National Oceanic and Atmospheric Administration
NRC	National Research Council
NSF	National Science Foundation
OAII	Ocean-Atmosphere-Ice Interactions
ODP	Ocean Drilling Program
ONR	Office of Naval Research
PAGES	Past Global Climate
PALE	Paleoclimates of Arctic Lakes and Estuaries
PICES	Pacific ICES
RIDGE	Ridge Inter-Disciplinary Global Experiments
ROV	Remotely Operated Vehicle
SCOR	Scientific Committee on Oceanic Research
SeaWIFS	Sea-Viewing, Wide Field Sensor
SERDP	Strategic Environmental Research and Development Projects
SOFAR	Sound Fixing and Ranging
SOSUS	Sound Surveillance Underwater System
SST	Sea-Surface Temperature
TAO	TOGA Tropical Atmosphere Ocean
TOGA	Tropical Ocean Global Atmosphere
TOGA COARE	TOGA Coupled Ocean-Atmosphere Response Experiment
TOGA SSG	TOGA Scientific Steering Group
TOPEX/Poseidon	Ocean Topography Experiment
T-POP	TOGA Program on Seasonal to Interannual Prediction
UNEP	United Nations Environment Program

USGS	United States Geological Survey
VOS	Voluntary Observing Ship
WCRP	World Climate Research Program
WOCE	World Ocean Circulation Experiment
WMO	World Meteorological Organization
XBT	Expendable Bathythermograph

APPENDIX II - Parameters of Importance for Understanding the Ocean's Role in Global Change[1]

PARAMETER	TOOLS	SOCIETAL BENEFIT
GENERAL ATMOSPHERE-OCEAN MEASUREMENTS		
wind air temperature humidity precipitation sea state	ships buoys satellites aircraft coastal stations	shipping; structures; search & rescue; offshore operations; meteorological models; weather forecasts; recreation
PROFILING ATMOSPHERE-OCEAN MEASUREMENTS		
balloon profiles acoustic profiles optical profiles	ships buoys	weather forecasts; air quality; air-sea chemical fluxes
SEA LEVEL		
sea level	tide gauges bottom pressure guages	tides; storm surge; navigation; coastal zone management; coastal development; salt water intrusion; climate change; recreation
CURRENTS		
currents	ships buoys drifters	navigation; search and rescue; emergency response; beach erosion; sediment and pollutant transport; global circulation models

[1] Adapted from a table in *First Steps Toward a U.S. GOOS* (see Appendix III)

PARAMETER	TOOLS	SOCIETAL BENEFIT
TEMPERATURE		
sea-surface temperature subsurface temperature	ships buoys VOS	climate change; coastal meteorology; environmental quality; sea-level change; fisheries
SALINITY		
surface salinity deep salinity	ships buoys	ecosystem health; fisheries; environmental quality; recreation coastal ocean models
ICE		
ice edge and extent coastal observations	satellites aircraft	navigation; climate change; meteorology; climate models; ecosystem health; fisheries
SHORT AND LONG WAVE RADIATION		
aerosols gas exchange chemical precipitation photosynthetically available radiation	satellites ground stations ships	climate change; water quality; ecosystems health; status and trends; fisheries
OPTICAL MEASUREMENTS		
pigment detritus clarity sediment load spectral properties	buoys ships satellites	water quality; ecosystem health; eutrophication; navigation; toxic blooms

PARAMETER	TOOLS	SOCIETAL BENEFIT
NUTRIENTS		
NO_3, PO_4, NH_3, NO_2, and O_2	ships buoys satellites	water quality; ecosystem health; eutrophication; fisheries; recreation; status and trends
PLANKTON		
phytoplankton fluorometric measurements species identification growth	ships buoys satellites biochemical and molecular techniques	environmental quality (blooms); ecosystem health; recreation; fisheries; global change
ZOOPLANKTON		
biomass abundance species identification	ships (incl. VOS) acoustics optics biochemical and molecular techniques	ecosystem health; fisheries; global change; fisheries
CHEMISTRY		
POC, DOC, PCO_2, DIC contaminants	ships buoys submersibles ROVs, AUVs	environmental quality; ecosystem health; living resources; recreation; human health
GEOLOGY		
sediment types grain size, porosity movement, morphology bathymetry-topography benthic biology coastal erosion-run off	cores grab samples echo sounder nets landers-incubators	coastal zone management; environmental quality; ecosystem health; navigation; living resources

APPENDIX III - Bibliography, Program Documents, and Addresses

Ocean Studies Board Contacts

William Merrell
Chairman, Ocean Studies Board
Texas A&M University
5007 Avenue U
Galveston, TX 77551
Tel: (409) 740-5732
Fax: (409) 740-7934
E-mail: W.MERRELL (Omnet)

Ocean Studies Board Office
HA-594
National Research Council
2001 Constitution Ave., NW
Washington, D.C. 20418
Tel: (202) 334-2714
Fax: (202) 334-2530
E-mail: M.KATSOUROS (Omnet)

General Publications

National Research Council. The Ocean's Role in Global Change: The Contemporary System—An Overview of Major Research Programs. National Academy Press, Washington, D.C. 1990. 10 pp.

National Research Council. Oceanography in the Next Decade: Building New Partnerships. National Academy Press, Washington, D.C., 1992, 202 pp.

Workshop on the Economic Impact of ENSO Forecasts on the American, Australian, and Asian Continents. August 11-13, 1992, The FSU Conference Center, The Florida State University, Tallahassee, Fla. 1993.

GOOS

Contact
Melbourne G. Briscoe
U.S. GOOS International Ad Hoc Working Group, *Chairman*
Office of Ocean and Earth Sciences
NOAA, National Ocean Service
1305 East-West Highway, SSMC4 Station 6616
Silver Spring, MD 20910
Tel: (301) 713-2981
Fax: (301) 713-4392

Publications

The Global Climate Observing System. A Proposal Prepared by an Ad Hoc Group, convened by the chairman of the Joint Scientific Committee for the World Climate Research Programme at Winchester, UK, 14-15 January, 1991. The Meteorological Office, Bracknell, UK, 1991. 21 pp.

Global Ocean Observing System. Resolution XVI-8 of the Sixteenth Session of the Intergovernmental Oceanographic Commission, Paris, 7-21 March, 1991, in: IOC-SC/MD/97.

Global Ocean Observing Systems Workshop Report. September 10-12, 1990, Alexandria, Va. 115 pp.

Toward a Global Ocean Observing System. D. James Baker, in: Oceanus, Vol 34, No. 1, Spring 1991, pp. 76-83. Woods Hole, MA.

A Global Ocean Observing System. Dana Kester, Gunnar Kullenberg, and Muriel Cole, in: Nature and Resources, Vol. 28, No. 1, 1992, pp. 26-34. UNESCO.

The Global Climate Observing System (GCOS): Responding to the Need for Climate Observations. An Introduction, April 1992, Joint Scientific and Technical Committee, GCOS. WMO No.777, 1992, Geneva. 12 pp plus covers.

First Steps Toward a U.S. GOOS. Report of a Workshop on Priorities for U.S. contributions to a Global Ocean Observing system. Woods Hole, Mass., 14-16 October 1992. 54 pp.

Global Ocean Observing System. Status Report on Existing ocean Elements and Related systems. IOC/INF-902, Paris, December 1992. 77 pp.

Interim Design for the Ocean Component of a Global Climate Observing System. Ocean Observing System Development Panel, February 1993, Texas A&M University, College Station, Tex. 105 pp.

The Approach to GOOS. Intergovernmental Oceanographic Commission, IOC-XVII/8 Annex 2 rev., Paris, 12 March 1993. 19 pp.

The Approach to the Global Ocean Observing System. Wolfgang Scherer, Albert Tolkachev, Gunnar Kullenberg, and Muriel Cole, in: WMO Bulletin, Vol 42, No. 2, April 1993, pp. 118-123. Geneva.

In Preparation:

U.S. National Report on Contributions to the Global Ocean Observing System. For presentation in April 1994 to the IOC Committee for GOOS, Melbourne, Australia.

Global Ocean Observing System. Special Issue of Oceanus, Fall 1994.

TOGA

Contacts

Kenneth Mooney
National Oceanic and Atmospheric Administration
Office of Global Programs
1100 Wayne Ave., Suite 1225
Silver Spring, MD 20910
Tel: (301) 427-2089
Fax: (301) 427-2082
E-mail: K.MOONEY (Omnet)

Edward Sarachik
Department of Atmospheric Sciences, AK-40
University of Washington
Seattle, WA 98195
Tel: (206) 543-6720
Fax: (206) 685-3397
E-mail: E.SARACHIK (Omnet)

Publications

National Research Council. El Niño and the Southern Oscillation: A Scientific Plan. National Academy Press, Washington, D.C. 1983. 72 pp.

World Climate Research Program. Scientific Plan for the Tropical Ocean-Global Atmosphere Programme. WCRP Publication Series No. 3. 1985

National Research Council. U.S. Participation in the TOGA Program: A Research Strategy. National Academy Press, Washington, D.C. 1986. 213 pp.

U.S. TOGA/COARE Science Working Group. TOGA/COARE Science Plan. 1989.

U.S. TOGA. Workshop Report: Ocean Observing System: Midlife Progress Review and Recommendations for Continuation. Nova University Press, Ft. Lauderdale, Fla. October 1989.

National Research Council. TOGA: A Review of Progress and Future Opportunities. 66 pp. 1990.

World Climate Research Program. International TOGA Scientific Conference Proceedings. Invited papers from the conference held in Honolulu, Hawaii, July 1990. World Climate Research Program, WCRP-43 (WMO/TD-No. 379). 1990.

World Climate Research Programme. CLIVAR: A Study of Climate Variability and Predictability. 33 pp. 1992.

National Research Council. Ocean Atmosphere Observations Supporting Short-Term Climate Prediction. National Academy Press, Washington, D.C. 1994.

National Research Council. Report by the Climate Research Committee in preparation.

WOCE

Contacts

Piers Chapman
U.S. WOCE Director
U.S. WOCE Office
TAMU/OCGN
College Station, TX 77843-3146
Tel: 409-845-8194
Fax: 409-845-0888

Eric J. Lindstrom
Program Scientist
U.S. WOCE Interagency Office
1825 I St. N.W.
Suite 400
Washington, D.C. 20006
Tel: 202-429-2039
Fax: 202-857-5219

Worth D. Nowlin, Jr.
Science Steering Committee, *Co-Chairperson*
U.S. WOCE Office
TAMU/OCGN
College Station, TX 77843-3146
Tel: 409-845-3720
Fax: 409-847-8879

Publications

U.S. WOCE, U.S. Contribution to WOCE Core Project 1: The Program Design for the Indian Ocean, 80 pp., U.S. WOCE Office, College Station, Tex., 1993.

U.S. WOCE Implementation Plan 1992, U.S. WOCE Implementation Record No. 4, 110 pp., U.S. WOCE Office, College Station, Tex., 1992.

U.S. WOCE Implementation Plan 1993, U.S. WOCE Implementation Record No. 5, 146 pp., U.S. WOCE Office, College Station, Tex., 1993.

JGOFS

Contacts

Hugh D. Livingston
U.S. JGOFS Planning Office
Woods Hole Oceanographic Institution
Woods Hole, MA 02543
Tel: (508) 457-2000 x2454
E-mail: hlivingston@whoi.edu (Internet)
H.LIVINGSTON (Omnet)

Otis B. Brown
U.S. JGOFS Steering Committee, *Chairman*
R.S.M.A.S., University of Miami
4600 Rickenbacker Causeway
Miami, FL 33149
Tel: (305) 361-4018
E-mail: obrown@rsmas.miami.edu (Internet)
O.BROWN (Omnet)

Publications

U.S. JGOFS Planning Reports 1-17.

Hawaii Ocean Time-Series Data Reports 1-3.

Bermuda Atlantic Time-Series Data Reports B-1 to B-3.

Bermuda Atlantic Time-Series Methods Manual, March 1993.

U.S. JGOFS Equatorial Pacific Process Study Protocols, 1993.

U.S. JGOFS, Livingston, H. D., and M. C. Bowles, Oceanus 35, 57-9, 1992.

Update: Joint Global Ocean Flux Study, Sea Technology, 49-53, Jan. 1993.

GLOBEC

Contacts

Thomas Powell
Division of Environmental Studies
University of California, Davis
Davis, CA 95616
Tel: (916) 752-1180
E-mail: T.POWELL (Omnet)

Brian Rothschild
University of Maryland
Center for Environmental and Estuarine Studies
Chesapeake Biological Laboratory
P.O. Box 38
Solomons, MD 20688
Tel: (410) 326-4281
Fax: (410) 326-6987
E-mail: B.ROTHSCHILD (Omnet)

Publications

Towards the Development of the GLOBEC Core Program (GCP). A report of the first international GLOBEC planning meeting. Ravello, Italy, March 31-April 2, 1992.

Population Dynamics and Physical Variability. Report of the first meeting of an international GLOBEC working group. Cambridge, England, February 1-5, 1993.

In preparation:

Sampling and Observation Systems. Report of the first meeting of an international GLOBEC working group. Paris, France, April, 1993.

Report of the ICES/GLOBEC Cod and Climate Change Working Group Meeting. Lowestoft, England, June 7-11, 1993.

GLOBEC-International Southern Ocean Working Group: report of the first meeting. Norfolk, Virginia, June 15-17, 1993.

Numerical Modelling. Report of the first meeting of an international GLOBEC working group. Villefrance, France, July 12-16, 1993.

ACCP

Contact

Kirk Bryan
Atmosphere and Ocean Sciences Program
Sayre Hall
Princeton, University
Princeton, NJ 08540
Tel: (609) 258-6571
Fax: (609) 987-5063

Publications

Bryan, K., and R. J. Stouffer. 1991. A note on Bjerknes' hypothesis for North Atlantic variability. Journal of Marine Systems 1:229-241.

Gordon, A. L., S. E. Zebiac, and K. Bryan. 1992. Climate variability and the Atlantic Ocean. EOS, Transactions of the AGU. 73(15):164-165

Delworth, T., S. Manabe, and R. J. Stouffer. 1993. Interdecadal variations of the thermohaline circulation in a coupled ocean-atmosphere model. J. Climate 6:1993-2011.

ATOC

Contact

Walter Munk
ATOC Project Office
University of California, San Diego
Institute of Geophysics and Planetary Physics
Scripps Institution of Oceanography
La Jolla, CA 92093-0225

Publications

The following papers were published as part of the Ocean '93 Conference (Engineering in Harmony with the Ocean, Proceedings Vol. I) held October 18-21, 1993, in Victoria, British Columbia, Canada.

ATOC Network Definition, D.W. Hyde.

Designing the ATOC Global Array, B. M., Howe, S. W. Leach, J. A. Mercer and R. I. Odum.

Acoustic Thermography for Arctic Ocean Climate, P. N. Mikhalevsky and R. Muench.

Numerical Solution of the Acoustic Wave Equation Using Chebyshev Polynomials with Application to Global Acoustics, M. Dzieciuch.

A Review of Ocean Current and Vorticity Measurements Using Long-Range Reciprocal Acoustic Transmissions, B. D. Dushaw, D. B. Chester, and P. F. Worcester.

GOALS

Contacts

Kenneth Mooney
National Oceanic and Atmospheric Administration
Office of Global Programs
1100 Wayne Ave., Suite 1225
Silver Spring, MD 20910
Tel: (301) 427-2089

Jay Fein
National Science Foundation
Division of Atmospheric Sciences, Rm 775
4201 Wilson Blvd.
Arlington, VA 22230
Tel: (703) 306-1527

Eric J. Barron
NRC Climate Research Committee, *Chairman*
Earth System Science Center
537 Deike Building
Pennsylvania State University
University Park, PA 16802
Tel: (814) 865-1619
Fax: (814) 865-3191

Publications

National Research Council. Report by the Climate Research Committee in preparation

LOICZ

Contacts

Scott Nixon
Graduate School of Oceanography
University of Rhode Island
Kingston, RI 02881
Tel: (401) 792-6800
Fax: (401) 789-8340
E-mail: S.NIXON (Omnet)

LOICZ Core Project Office
Netherlands Institute of Sea Research
PO Box 59
1790 AB Den Burg
Texel, the Netherlands
Tel: 31-2220-69404
Fax: 31-2220-69430

Publications

Land-Ocean Interactions in the Coastal Zone Science Plan. IGBP Report No. 25. The International Geosphere-Biosphere Programme: A Study of Global Change (IGBP) of the International Council of Scientific Unions (ICSU). Stockholm, February 1993.

ARCSS

Contacts

Until new ARCSS Executive Committee Chairperson is appointed, contact:

Arctic Research Consortium of the United States
600 University Avenue
Suite B
Fairbanks, AK 99709
Tel: (907) 474-1600
Fax: (907) 474-1604

Patrick Webber
National Science Foundation
Office of Polar Programs
4201 Wilson Blvd., Rm. 755
Arlington, VA 22230
Tel: (703) 306-1030
Fax: (703) 306-0139

Publications

Arctic Social Science: An Agenda for Action. Committee on Arctic Social Sciences, Polar Research Board, National Research Council. National Academy Press, Washington, D.C., 1989, 75 pp.

Arctic System Science: Advancing the Scientific Basis for Predicting Global Change. ARCUS, Boulder, Colorado, 1990, 8 pp.

Arctic System Science—Land Atmosphere/Ice Interactions Science Plan. ARCUS, Fairbanks, Alaska, 1991, 23 pp.

Arctic System Science—Ocean/Atmosphere/Ice Interactions, Initial Science Plan, Joint Oceanographic Institutions, Inc., Washington, DC, Report No.2, 1992, 27 pp.

Arctic System Science—Ocean/Atmosphere/Ice Interactions, SHEBA: A Research Program on the Surface Heat Budget of the Arctic Ocean, OAII Science Management Office, University of Washington, Seattle, Washington, Report No. 3, 1993, 34 pp.

Arctic System Science:A Plan for Integration. Arctic Research Consortium of the United States, Fairbanks, Alaska. 1993.

ODP

Contacts

Joint Oceanographic Institutions Inc.
1755 Massachusetts Avenue, N.W.
Suite 800
Washington, D.C. 20036-2102
Tel: (202) 232-3900
Fax: (202) 232-8203
E-mail: JOI.INC (Omnet)

Tim Francis
Ocean Drilling Program
1000 Discovery Drive
College Station, TX 77845-9547
Tel: (409) 845-2673
Fax: (409) 845-4857
E-mail: fabiola@nelson.tamu.edu (Internet)

Brian Lewis
JOIDES Office
University of Washington, HA-30
Seattle, WA 98195
Tel: (206) 543-2203
Fax: (206) 685-7652
E-mail: blewis@ocean.washington.edu (Internet)

Publications
The following publications are available from the Ocean Drilling Program Office.

Proceedings of the Ocean Drilling Program. Initial Reports, Volumes 101-145, and Scientific Results, Volumes 101-131.

Scientific Prospectuses. Volumes 1-54.

Preliminary Reports. Volumes 1-51.

Technical Notes. No. 1,3,6-18, 20-22.

RIDGE

Contact
Robert Detrick
Woods Hole Oceanographic Institution
Woods Hole, MA 02543
Tel: (508) 457-2000 x3335
Fax: (508) 457-2187
E-mail: R.DETRICK (Omnet)

Publications
RIDGE Science Plan 1993-1997. RIDGE Office, Woods Hole Oceanographic Institution, Woods Hole, Mass., 1992, 101 pp.

USGS Global Change

Contact
Richard Poore
U.S. Geological Survey
National Center MS 955
12201 Sunrise Valley Drive
Reston, VA 22092
Tel: (703) 648-5270
Fax: (703) 648-6647

Publications

Report of a workshop on the correlation of marine and terrestrial records of climate changes in the western United States. J. V. Gardner, A. M. Sarna-Wojcicki, D. P. Adam, W. E. Dean, J. P. Bradbury, and H. J. Rieck. U.S. Geological Survey Open-File Report 91-140, 1991, 48pp.

Proceedings of the U.S. Geological Survey Global Change Research Forum, Herndon, Virginia, March 18-20, 1991. John A. Kelmelis and Mitchell Snow, eds., U.S. Geological Survey Circular 1086, 1993, 121 pp.

Poore, Richard Z. Editorial, Paleoceanography, v. 8, pp. 135-136. 1993.

Grantz, Arthur. Cruise to the Chuchi Borderland, Arctic Ocean. EOS Transactions, American Geophysical Union, v. 74, pp.249, 253-254, 1993.

MESH

Contacts

Nicklas G. Pisias
College of Oceanic and Atmospheric Sciences
Oregon State University
Ocean Admin. Bldg 104
Corvallis, OR 97331-5503
Tel: (503) 737-5213
Fax: (503) 737-2064
E-mail: N.Pisias (Omnet)
pisias@oce.orst.edu (Internet)

Bilal Haq
National Science Foundation
Ocean Sciences Research Division, Rm 725
4201 Wilson Blvd.
Arlington, VA 22230
Tel: (703) 306-1586
Fax: (703) 306-0390
E-mail: B.HAQ (Omnet)
bhaq@nsf.edu

Publications

Advisory Panel Report on EARTH SYSTEM HISTORY. An Element of the U.S. Global Change Research Program, Joint Oceanographic Institutions, Inc. 1991.

IMAGES—International Marine Global Change Study. A marine component to PAGES pilot program PANASH II. Proposal for joint SCOR/IGBP-PAGES. In Press.

MESH Program Plan. In Preparation.